About Island Press

Island Press, a nonprofit organization, publishes, markets, and distributes the most advanced thinking on the conservation of our natural resources—books about soil, land, water, forests, wildlife, and hazardous and toxic wastes. These books are practical tools used by public officials, business and industry leaders, natural resource managers, and concerned citizens working to solve both local and global resource problems.

Founded in 1978, Island Press reorganized in 1984 to meet the increasing demand for substantive books on all resource-related issues. Island Press publishes and distributes under its own imprint and offers these services to other nonprofit organizations.

Support for Island Press is provided by The Geraldine R. Dodge Foundation, The Energy Foundation, The Ford Foundation, The George Gund Foundation, William and Flora Hewlett Foundation, The James Irvine Foundation, The John D. and Catherine T. MacArthur Foundation, The Andrew W. Mellon Foundation, The Joyce Mertz-Gilmore Foundation, The New-Land Foundation, The Pew Charitable Trusts, The Rockefeller Brothers Fund, The Tides Foundation, Turner Foundation, Inc., The Rockefeller Philanthropic Collaborative, Inc., and individual donors.

MORE TREE TALK

"If you spend your life out in the woods, you get to appreciate nature's way of doing things. When I'm falling timber and I get a little bit tired, I set down and sometimes I watch a bunch of little ants. There was a whole string of them; they were pulling these little fir seeds into this ant hi'l. Well, you know how an ant hill is; it's all fertilized. I was really curious about that, and I went back two years later and here was all these little trees growing out of there. I thought to myself, nature's got a system where they've got animals planting twenty-four hours a day. All the little squirrels, and the birds—nocturnal animals, too. How can man compete? People talk about how they're planting six trees for every old-growth; that can't even come close to what nature's doing, and it never will."

—*Wally Budden*

"It's a whole other way of doing forestry. We're not looking at a quarterly profit; we're looking at a very long-term sustainability through many generations of people and trees. The real brilliance of Jan's vision was tying the economic structure of community-based forestry into the definition of sustainable forestry. The economic part of it fits completely in with the ecological side. One doesn't happen without the other. You can't expect to have a stable economic base without ecological forestry, and vice versa. The two go hand-in-hand. The economic structure of the area is based on the stability of that resource. And the best way to insure sustainability is to have the decisions being made by the local people who have a real interest in the future of the forest. If we want to break the boom-and-bust cycle which is so common in rural areas such as ours, if we want to create jobs for the young people of the community, the people living in an area have to maintain control of their resource base."

—*Peggy Iris*

MORE TREE TALK

The People, Politics, and Economics of Timber

Ray Raphael

ISLAND PRESS
Washington, D.C. • Covelo, California

Illustrations by Mark Livingston

Library of Congress Cataloging-in-Publication Data
Raphael, Ray.
    More tree talk: the people, politics, and economics of timber/Ray Raphael.
      p. cm.
    Includes bibliographical references (p.   ) and index.
    ISBN 1-55963-253-4 (cloth).—ISBN 1-55963-254-2 (paper)
    1. Forests and forestry—Northwest, Pacific.   2. Forest management—Northwest, Pacific.   3. Forests and forestry—Economic aspects—Northwest, Pacific.
4. Forest ecology—Northwest, Pacific.   5. Forests and forestry—Social aspects—Northwest, Pacific.   6. Forest policy—Northwest, Pacific.   7. Timber—Northwest, Pacific.   I. Title.
SD144.A13R36   1994
333.75'0973—dc20   93-48896
                                                                CIP

Printed on recycled, acid-free paper

Manufactured in the United States of America

10   9   8   7   6   5   4   3   2

For Nick, Neil, and
Marie Raphael, and in
fond memory of
Jim Demulling

Contents

PART III
Money in the Forest
Who Owns the Land?

Acknowledgments

First, I'd like to thank the people interviewed for both editions of *Tree Talk*. Without their gracious participation, this book could not exist in its present form. The wide variety of perspectives which they offer adds to the book's vitality. Yet since their viewpoints are varied, nobody is to be held responsible for the words that others might say elsewhere in the book, nor does a personal appearance in *Tree Talk* constitute an endorsement of my own statements and conclusions. Each person speaks only for himself or herself.

In addition to those who offered valuable information and assistance for the first edition, I would like to thank the following people who gave interviews, answered questions, helped guide my research, or reviewed portions of the manuscript: Michael Evenson, David Simpson, Luis Torres, Jeff Romm, Sungnome Madrone, Jim Rinehart, Rich Fairbanks, Seth Zuckerman, Brett KenCairn, Helen McKenna, Rondal Snodgrass, Bill Stewart, Don Gasser, Ed Stone, William McKillop, Ben Twight, Rudolf Becking, Bill Wilkinson, Jeff Stier, Jeff DeBonis, Mike Geniella, Deena Kraft, Roy Keene, Bob Hrubes, DeRay Norton. I would also like to thank Tony Mengual, John Rogers, Gil Gregori, Bill Eastwood, Tracy Katelman, Jerry Wilson, Dave Kahan, and the other committed people at the Institute for Sustainable Forestry for turning ideas into action, and all the tireless workers involved in Vision 20/20, the Rogue Institute, the Applegate Partnership, Forest Trust, and a host of other projects that are seeking pathways to a sustainable future.

Introduction

The old-growth forests are almost gone. Loggers are out of work. Wildlife species such as the spotted owl are seriously threatened. With the weakening of economic activity in timber-dependent communities and a corresponding decline in the health of forest ecosystems, the politics of timber have become increasingly volatile. According to some, the timber industry is responsible for the rape of a treasured resource; for others, it is the environmentalists who are to blame for making self-respecting workers stand in line for food stamps. While each side villifies its opponents, the fabric of life (both human and otherwise) has forever changed in rural communities which were once perceived as idyllic.

Now is the time for healing. If we wish to move beyond the current impass, we must address both economic and ecological concerns. The basic economic argument cannot be denied: Human beings struggling to survive on a finite planet are natually going to make use of the available resources. But the ecological argument cannot be denied either: Maintaining healthy ecosystems is important both for the sake of the forests themselves and to insure an adequate resource base for the future. To provide for our lasting needs, we must learn to treat the forests respectfully.

When I wrote the first edition of *Tree Talk* thirteen years ago, I hoped to transcend the political polarization that had already begun to cripple our timbering regions. In order to shake up old myths and stimulate new lines of reasoning, I structured the book around interviews with men and women of diverse persuasions who had direct knowledge of the woods: cat loggers and horse loggers, naturalists and timber company executives, old-time woodsmen and youthful restoration workers. The book took the form of a documentary film: the running narrative was punctuated by individual portraits intended to personalize the issues, to translate both the political and academic aspects of forestry into human terms.

In *Tree Talk,* I postulated a middle road between overmanagement and no management at all. I called this approach *holistic forestry:* human intervention, if neccesary, was to complement rather than contradict natural forces. Forestry, I reasoned, should be based on the principle of maintaining a balanced ecosystem—not just in a handful of old-growth preserves, but throughout our commercial timberland. The task at hand was to figure out some way of providing the products we need on a sustainable basis—to learn how to extract timber from the woods while simultaneously preserving a stable environment.

In the past decade, this approach to forestry has moved from the fringe into the mainstream. *New forestry,* as it is now called, is accepted in official circles, even forming the basis of President Clinton's plan to resolve the spotted owl controversy. Its meteoric rise to official acceptance, however, is a mixed blessing. The term *new forestry* has managed to accrue some political baggage, being associated with specific programs that might or might not be popular. There is also something faddish about this nondescriptive terminology that runs counter to its serious intent. I have therefore chosen to stay with the term *holistic forestry,* even if it sounds a bit dated. At the very least, it retains some meaning in its name; *holistic* refers to whole systems, the overall picture. The intent is clear: the forest is to be perceived as a complex of interactions, not simply a bunch of trees that we wish to grow and cut as quickly as we can.

In the first edition, I admitted that holistic forestry was "only an idea, a direction, a consciousness." I concluded the book with a hypothetical interview of a "landed forester of the future"—largely because there was not much happening out on the ground. This time around, there is more positive news to report. Holistic forestry has begun to take shape in the woods. The Forest Service—and even many industrial foresters—now speak in terms of *ecosystem management,* with its emphasis on biological diversity, rather than *intensive management,* with its apparent bias toward timber. Many environmentalists, meanwhile, are turning into *forest practitioners,* trying to make their living by harvesting sustainable forest products. While heated debates continue to dominate the headlines, increasing numbers of well-intentioned people are beginning to find ways in which forest managers can use natural laws to sustain viable ecosystems—while still producing timber for human utilization.

In *More Tree Talk,* I have therefore been able to delete the hypothetical in favor of the real, adding 15 new interviews with people who have fresh perspectives, constructive ideas, and workable projects. All the subjects claim to accept the basic ideal of maintaining a healthy environment which can support a strong economy, although there are widely differing opinions as to how this goal might be achieved.

According to some, clearcutting can be consistent with sustainable forestry; according to others, uneven-aged management is the only way to go.

I have made two other major adjustments. First, I have reviewed the scientific literature, updating the research that has been done since 1981. Many of the significant advances in our understanding of forest ecosystems—the role of woody debris, the functioning of mycorrhizae, the importance of biological diversity—are reported in the first section of the book.

But scientific understanding is not enough. Despite our increased knowledge and the best intentions of forest practitioners, our actual treatment of the woods will not improve unless we can recognize—and change—the economic and political forces which interfere with good forest management. Forestry, like any other enterprise within a capitalist system, is market-driven. Yet the economics of forestry are poorly understood by most people, even by many who profess an interest in the fate of the forests. I have therefore paid particular attention to financial matters in this revised edition, even adding the word *economics* to the subtitle. I have also devoted separate chapters to the three forms of forest ownership: public, industrial, and private. My premise is that good, sustainable forestry will not be practiced until the structural obstacles inherent in each of these ownership forms are understood and addressed.

Economic issues are at the forefront in the Pacific Northwest, where the politics of timber have reached a fever pitch. Although *More Tree Talk* focuses on this specific region, many of the lessons to be learned here can quickly be adapted to other areas. The industrial tree farms of the Southeast, the private woodlots of the North and Northeast—these are all still forest ecosystems, they are all still subject to political regulation, they are all still driven by market realities. Although the individual communities might not be as dependent upon timber, I trust that people elsewhere who have a stake in their future forests can benefit by studying the much-magnified problems that we are experiencing in the Pacific Northwest.

The problems we are facing here are not that unique. As the human population of the world continues to increase, we place increased pressure on our finite resources. Unavoidably, we are having a more serious impact upon our environment, which in turn leads to a more serious concern about that impact. The struggle between resource extraction and environmental protection is basic to our times. The battles we fight will not be resolved overnight; indeed, they will never be resolved if we continue to perceive our human economy as separate and distinct from a global ecology. Here in the forests of the Pacific Northwest, we have an opportunity to become pioneers in the stewardship of a sustainable environment that can support a sustainable economy.

The Background of Debate

1

TREE MINING
The Voice of History

When European settlers first touched upon the eastern seaboard of North America, they were confronted by a thick and apparently endless forest. Centuries earlier, Europe itself had been covered with a similar forest; but an increasing population, which required an increasingly high standard of living, had transformed the deep woods of the Old World into farmland. Now, the colonists would attempt another conversion in their new home: razing the mysterious, wild forest to make room for fields and pastures. These people perceived the forest more as an obstacle to progress than as a precious resource.

With trees to spare, the settlers developed a technology that utilized wood for a variety of purposes. Houses, barns, ships, bridges, fences, carriages, furniture, tools, barrels—the appliances of daily life were fashioned from the trees that the settlers wished to clear from the land. Wood was burned in every home for cooking and heating; it was also turned into charcoal for the blacksmiths' forges. With the introduction of the steam engine in the eighteenth century, charcoal became the primary source of energy for industrial production. Throughout the seventeenth, eighteenth, and nineteenth centuries, more wood was consumed as fuel than was made into lumber in the United States.

By the middle of the nineteenth century, most of the desirable farmland east of the Mississippi had been cleared of trees. During the Civil War the remaining forests—mixed hardwoods situated on hilly terrain or infertile soil—were cut down for charcoal to fire blast furnaces for wartime production. By the time the industrial furnaces finally switched over to coal a few years later, the vast eastern woodlands had been totally consumed. Forests once thought inexhaustible had, in fact, been exhausted.

Forestry's age of innocence had ended: the timber resource could no longer be taken for granted. Lacking the timber for both fuel and lumber, several eastern states offered bonuses and tax incentives for the planting and nurturing of trees. But the forests could not grow back overnight, and the demand for wood products continued after the first generation of trees had been used up. Lumbering interests did not wait around for the better part of a century for new trees to mature. Instead, they moved on to where the trees still stood tall, to forests that had not yet been cut. During the second half of the nineteenth century, the pines of New England were superseded by bigger and better pines from the great North Woods of Michigan, Minnesota, and Wisconsin. But these, too, proved finite in number. As the century drew to a close, the best of the timber in the North Woods had been cut. Once again, the tillable land was turned into farms; on the areas unclaimed for field or pasture, another crop of trees would be a long time in the making.

Farther west, there were trees still bigger and better than any that had been seen before: the majestic redwoods of California and the giant Douglas-firs, cedars, spruces, and hemlocks of Oregon and Washington. The forests here were on an altogether different scale. Many of the trees stretched 250 feet into the sky and measured 50 feet around at the base of the trunk. A single tract of forested land could be as large as an entire New England state. Confronted with a logger's paradise, the western pioneers boasted that they had at last discovered an inexhaustible source of timber. "California will for centuries have virgin forests, perhaps to the end of Time," remarked an awe-struck admirer in the 1850s.[1] There was no talk in the Old West about trees being a renewable resource; if the virgin timber would last forever, the trees didn't have to be renewed. Timber was extracted from the surface of the earth in much the same way as minerals were mined from underneath the ground. There was no more thought of replacing a 250-foot tree than there was of returning gold dust to the streams.

The unsettled tracts of western timberland were so vast that the government had to develop special strategies for dispensing with them. Some fell into the hands of the railroads as part of the land-grant rights-of-way, but the avowed goal of the federal government was to disperse the ownership as widely as possible. To this end, the famous Homestead Act of 1862 was supplemented by the Timber Culture Act of 1873: a homesteader could add 160 acres to his claim simply by agreeing to grow trees—that is, not to cut them down—on 40 of the extra acres that the government gave him. The Timber and Stone Act of 1878 enabled citizens of the West Coast states to purchase 160 acres of public timberland, for personal use only, at the nominal charge of $2.50 per acre. In this manner the spoils of the ubiquitous timber were supposed to be distributed equally to a democratic army of independent loggers, the West Coast equivalent of the small, independent farmers of the East and Midwest.

It didn't work out that way. In practice, the laws were easily manipulated to serve the purposes of more organized logging interests. Companies employed individuals to purchase public timberlands at the nominal fee and then turn over the deed. Homesteaders who took up the 160-acre timber preemption satisfied the government by preserving 40 acres of trees, but then quickly sold off the timber on the remaining 120 acres to whichever company happened to be operating in the area. The trees were thus made available to those who could most readily utilize them: the local logging empires that were emerging up and down the Pacific Coast.

Only in rugged terrain or inaccessible localities did the democratic distribution of timber work as intended. Far from railroads or rivers, homesteaders in the backwoods harvested products of the forest primarily for use by themselves or their close neighbors. Occasionally they processed the trees into small hand-split products that could be transported out of the woods on the back of a mule: shakes, fence posts, railroad ties, grapestakes. But the logs had to be processed right in the middle of the forest. The mules could not carry uncut logs to the mills, so full-scale lumbering remained impractical in the backwoods.

For the most part, the ideal of settled, self-employed loggers could not be realized in the West. Nothing short of a fully organized crew working in a coordinated effort would suffice to move the giant timber out of the woods. In the eastern and Great Lakes states, a homesteader had only to wait for winter if he wanted to log by himself: he then placed his logs on sleds and had his horses slide them out of the woods over the snow and ice. In the West, though, the trees were too large and the land too uneven for such a one-man operation, and there often was no snow or ice in the winter. Here, logging required a new assortment of techniques utilizing a unique combination of gravity, water, animal power—and human sweat.

Big Tree Technology

The first problem faced by a western lumberjack, strangely enough, was to climb the tree to its trunk. At ground level the swell in a giant redwood or cedar might be eighteen or twenty feet in diameter, while ten feet up the trunk would shrivel to a mere twelve or fourteen feet. To save work and avoid the pitch in the stump, the fallers climbed the tree by chopping a small hole through the bark and inserting a springboard, a flexible plank. Often, three or four springboards would be used as a ladder to reach the desired height. Working from these portable scaffolds, a pair of fallers could chop down the tree with axes and two-man saws, a task that might take a day or two for each tree.

Once the tree lay safely on the ground, the buckers removed the limbs and cut the trunk into logs. With hand-jacks and peavies, the lumberjacks rolled the logs around stumps and down the sidehills into the gullies. There, teams of oxen or horses would drag the logs to the main skid roads. Since there was no snow or ice upon which the logs might slide, roads were constructed of solid rows of timbers which had to be watered or greased to minimize friction. The bull-puncher or bull-whacker (his name varied with the location) pounded dogs (hooks) into the end of each log, connected the logs to his team with a rope, and drove the team down the skid road to the nearest mill or river.

Skid roads were seldom more than a mile or two in length, for overland travel was difficult and tedious. In the early days, the preferred means of transportation over long distances was by water. The presence of any type of water—oceans, rivers, and even streams—facilitated the movement of logs. Along the Pacific Coast, logging schooners anchored

in "dog-hole" ports to take on their cargo. Sometimes primitive wharves were constructed; more often, the logs were loaded by cables suspended from the nearby bluffs. Inland, the logs were dumped into the still backwaters of the rivers during the dry summer months. If several outfits operated along the same river, logs were branded with the insignias of the owners and sorted out down at the mills. When the high water came at the end of the season, the logs tumbled downstream. If a log got hung up on a shallow riffle, the "river rats" pried it loose with their peavies. If all else failed, the troublesome log was hitched to an animal team on shore and dragged over the riffle to deep water.

In most areas, it was only a matter of a decade or so before the forests immediately adjacent to the rivers had been exhausted. To reach farther into the hinterlands, the loggers had to utilize smaller and smaller streams. In order to trap enough water in these tributaries to transport the logs, the loggers built splash dams upstream from the decked timber. When released, these dams created man-made floods that literally flushed the logs down to the mills.

Soon even the timber along the streambanks was stripped away. Millions of acres of untouched forest remained throughout the West, but the trees could not be reached by water. Full-scale harvesting in the

The ox team

backcountry awaited the development of a mechanized overland technology that could transport giant logs over extended stretches of rugged terrain.

Sid Stiers

Lumberjack and Saw Filer

"My grandparents come out here in the fifties. On my mother's side, eighteen and fifty. My dad's folks come in about eighteen and fifty-six or seven. Worked in the woods—that was all there was in them days.

"They had ox teams at about that time, and then it wasn't long 'til they got horses. Team of good horses will outpull a team of good oxen. Faster. Them old oxen was slow, you know. They'd just mope along there. But when they come out here, you see, when my grandparents come across the plains, that's all they had was oxen. The horses was zero. In the East they had a few horses, but out West was empty country. Finally they got to bringing horses in, see.

"I was seven years old. My dad was logging up here on our old homestead. Had quite a lot of timber there. He logged it to go on the river drive. I run the old water sled. I was big enough. I didn't have to know much, the old horse knew more than I did. Just had a barrel on a sled, and these skids was so far apart. Two runners on it, planks across it, and a big old fifty-gallon barrel set on that. You'd get her started out ahead of a string of logs—gosh, I don't know how many logs. After they got them on the skid road, they'd trail them. So I'd go along with that and pull the pin in the barrel, and that water'd string out along there and help them skid. A lot of them used grease—go along with a bucket of grease and a mop ahead of a big turn of logs. Those logs, if you didn't have the skids watered or greased or something like that, they'd just pull harder than heck. Made worlds of difference.

"The old river drives were about turn of the century, along in there. All the men would dump the logs into the river 'til they got their logging done. They'd get together, all of these outfits that was dumping in the river, and they'd get a big log drive a-goin'. Took quite a time. They'd start in the fall. Maybe it'd rain a little and raise the water, and then they'd take off. They'd end up way late at the old millrace down there at Eugene, where it takes out of the river. Big logjams. They'd have to pull the logs out to get them going.

Peavies poised, turn-of-the-century Washington "river rats" release logs back into the current

Farther down the river, all the logs would get hung up on a riffle. Just one riffle after another, they'd have to pull the logs off.

"They'd get the logs into the millrace, and it'd be freezing. They had to wade out and drive the dogs in the logs, and then they had a guy who rode the logs out. They'd get on a log where it's about waist deep and off they'd go, hollering all the way. Cold! But once they got wet it wasn't so bad. They'd get wet anyway, so a lot of them would just jump in and get it over with.

"One time they was logging sugar pine way up the Willamette. So they logged that sugar pine and put it in the river. They even built a dam. The Willamette up there isn't too big, so they built a flood dam. They got a few logs in and they turned the water on and by-golly—nothing. The logs just laid there. They'd sink, see, just go down like a rock. Sugar pine don't float. I don't know why they didn't know they'd sink. They had to go down there with hooks and fish them all out.

"I was too young to work on the river drives. I was just a pest along about then. Later ones, I was probably about twelve. First job I really went out on, I was sixteen years old. I was bucking wood for an old Willamette swing donkey. I worked in the woods, on the rigging or anything, 'til nineteen and seventeen when I got a job filing the old hand briars, the misery whips. From then on I did some falling and bucking, you know, but mostly my job was filing old hand brim. That was my trade for about thirty-five years.

"Worked for lots of different outfits. Lived in logging camps. Dirt floors and straw mattresses. It was a mess, you look at it now, but we didn't think much about it in them days. One place, they had just a little old rake and they'd rake it out every once in a while, when you couldn't get in. Those old beds were just made out of boards: poles up and a bottom on it, and you'd fill that with straw. When it got too hard, you throwed that one out and built yourself another bed of hay. Had to carry your own bedroll. They'd have a row of bunks clear around the wall, two bunks high, one at the bottom and one at the top. And benches to set on, and that's all they had. Set there to put your cork shoes on.

"The old cookhouses, they fed good. But, boy, you get a big crew sitting down to the table, and you had to be fast or you wouldn't get much to eat. Fellow had to raise up and reach clear over the table, if he wanted to get a good hunk of meat. By golly, it was quite a deal. They had good cooks; mostly women cooked in those camps. Just common meat, potatoes, and vegetables when they got them, and they'd bake pies and stuff like that. Lot better than you get now. That come from the ground up, what they cooked. If they didn't put out the food, well: 'Ain't nothing to eat here, I'm going someplace else.' And away they'd go.

"It was rough living in them camps. West Fir was one of them in the twenties. To get a place to live I lived in a tent. All I had was a little, wood, two-door cookstove. Tough going. If you wanted to work, you had to put up with it until something come along better.

"Booth-Kelly, they was a bigger outfit. They had these houses built on skids. So they'd move them on a flatcar, and then when they got to where they was going to have a camp, they'd slide the houses off. When they got to working quite a ways off in some other direction, why they'd slide the houses back on and move to another camp.

"You know, there was always a joker around one of them camps. Worse then than it ever is now. This one guy he got up early and built this fire up good and hot and then wet on it and away he went. Man, I tell you the covers was a-flyin'. But they got even with him somehow. Son-of-a-gun, there was always someone pulling something on somebody.

——— " ———

Pranks may have occurred, but the logging camps were not noted for high living. In most camps, rowdy behavior was cause enough to be fired, as was socializing with the cooks or waitresses. The days were taken by work, and at night silence was enforced in the bunkhouses, in order that the men might sleep. In some camps, a strict silence was even enforced at mealtime.

The inhibitions of camp life, however, were relieved by periodic trips to the nearest city or town. Because of the seasonal nature of employment, nearly every lumberjack spent a good part of the year out of the woods. Near the mills on the outskirts of town, shanties, cheap hotels, saloons, and restaurants sprang up to care for the transient workers. These sections became known as "skid rows," named after the old skid roads upon which they were located. Life on skid row was the reverse image of life in a logging camp: there was no work to be had, no restrictions on personal behavior, but plenty of vices upon which money might be spent.

Seasonal migrations of the workers were paralleled by the frequent relocation of the logging camps themselves. As each new section of the woods was logged out, the center of operations would move on. Such an unstable situation was not conducive to a settled life among the loggers. More common than the family man was the "boomer," or tramp logger, who followed a loosely plotted trail from camp to camp. According to Rex Seits of Lacomb, Oregon: "The old tramp logger used to, well, he liked to see the other side of the hill once in a while and learn new games, you know, different ways. That's their life: just to keep moving; although they're just out hunting for a good cookhouse. And your companies at one time would just as soon you did, because when a new man come in, why, they'd get a lot more work out of him. He was faster, quicker, energetic; he was always in a hurry. After he was there a while, maybe he'd slow down."[2] Jess Larison, a timber faller from Dexter, Oregon, recalls: "I spent several years in those camps, and I never saw a single one of those men again."[3]

The mark of the tramp logger was a bedroll on his back; in self-mockery, he called himself an "A. P. A." (American Pack Animal). As often as not, his bedroll was infested with fleas or lice, which were easily spread from bunkhouse to bunkhouse. When the radical Industrial Workers of the World (IWW) organized in the woods just before and after World War I, their rallying cry struck home in the hearts of the lumberjacks: clean bedding should be supplied by the companies at all logging camps. The call for clean bedding probably won more converts for the Wobblies than did the demand for higher wages.

Picking Up Steam

By the time the workers in the woods became organized, mechanization had altered the fundamental nature of their occupation. Two types of steam engines—railroad locomotives and high-powered winches called steam donkeys—had revolutionized the transportation of logs. Railroad tracks could be laid where water never flowed, while steam donkeys could pull logs both faster and easier than the cumbersome animal teams. The steam era ushered in the famous "highball" logging shows, in which production and efficiency surpassed all previous hopes and expectations. Unlike the old river drives, the railroad operated year-round; unlike the old animal teams, the steam donkeys were mindless machines that had neither independent wills nor stomachs to feed. Indeed, the only "food" required by the railroad locomotive and steam donkey was wood—and there was plenty of that around.

The first steam donkey was put into use in 1881 in northwestern California by John Dolbeer, who took out a patent on his invention two years later. The design was simple: a steam-powered engine reeled a cable onto a spool, with a log attached to the end of the line. The engine, anchored to a sled, could pull itself along a skid road with its own power by use of a block and tackle hung on a nearby tree or stump. Logs could therefore be yarded from any position along the skid road.

The early donkey engines worked side by side with animals. The steam donkeys yarded the logs from the woods to the skid roads, where the animals took over. After each load was yarded, a horse—or a man—had to drag the cable back into the woods so it could be secured to another log. Gradually, however, the techniques of donkey-cable logging were streamlined and perfected. With the addition of a separate spool to the engine, an extra haulback line was able to lead the mainline back into the woods after each turn was completed, eliminating the need for men or animals to carry around the cable. And during the 1890s, a more powerful machine called a bull donkey began to replace the animal teams on the skid roads.

In the early systems, the logs were dragged along the ground where they encountered excessive friction, ran into obstacles, and carved deep trenches in the earth. By hanging the mainline from a stump, however, the front end of the log could be lifted slightly off the ground. Through the years, stump rigs climbed higher and higher, until they were 200 feet above the ground on specially prepared spar trees. From a single spar tree on a landing, a high-lead system could stretch out in a fan-shaped configuration and log everything within a radius as long as the mainline. In the most sophisticated forms of cable logging, both ends of a line were suspended from trees, enabling the entire log to be lifted off the ground as it was yarded.

The impact of the steam engine on the woods was profound: more trees could be cut from more distant locations. Terminal logging railroads could be built in the vicinity of any prospective mill site to provide major arteries into the forest. When operations were finished for a given site, both the mill and the tracks would be taken up and reassembled amidst an unharvested tract of land. Reaching out from these short, specialized railroad lines, the bull donkeys provided feeder systems to the right and left, and these in turn were fed by the yarding engines, which extended their cables directly to the logs on the ground. Since yarding with steam donkeys was much more efficient, once-marginal lands could now be profitably logged.

Clearcutting became a practical necessity in donkey logging, for the cables needed room to maneuver without running into standing trees. The railroads, too, led indirectly to clearcutting: the capital costs of setting up shop at any particular location required the company to take all it could get before moving on. No longer were only the best and most accessible trees taken; anything that could pay its way out of the woods wound up at the mill. As for trees that could not pay their way, they were either burned on the spot as impediments or burned as fuel for the donkeys and locomotives. Indeed, the railroads themselves became significant consumers of timber for ties, tunnels, and trestles.

As the extent and scope of the cutting increased, forward-looking citizens began to fear that the expansive forests of the West would follow the fate of the eastern woodlands and soon become exhausted. In the words of Gifford Pinchot, Forest Service director in the early 1900s, "The United States has already crossed the verge of a timber famine so severe that the blighting effects will be felt in every household in the land."[4] As a direct consequence of accelerated logging, a strong conservation movement developed around the turn of the century to push for the protection of timber resources. But "protection" at that time did not entail preservation as wilderness; the conservationists only wished to ensure that there would be trees to cut down in the future. In 1897 the Organic Administration Act set the goal for the newly formed National Forests: "to furnish a continuous supply of timber for the use and necessities of citizens of the United States."[5]

The interest in sustained-yield forestry was not limited to public lands. In 1908 Teddy Roosevelt convened a White House conference that included governors, members of the Cabinet and Supreme Court,

Mining the forest

scientists, and active conservationists. The conference adopted a "Declaration of Principles":

> We urge the continuation and extension of forest policies adapted to secure the husbanding and renewal of our diminishing timber supply, the prevention of soil erosion, the protection of the headwaters, the maintenance of the purity and navigability of our streams. We recognize that the private ownership of forest land entails responsibilities in the interests of all the people, and we favor the enactment of laws looking to the protection and replacement of privately owned forests.[6]

These principles, however, were never backed up by federal legislation, and a third of a century passed before any of the western states took such concerns seriously. The writing was on the wall, but for the time being the old-growth trees were still too numerous to count.

The effects of steam power were also felt on social and economic levels. Because of the high capital costs in railroad–donkey logging, small operations found themselves struggling under a distinct disadvantage. Logging concerns lacking the money to finance railroad tracks and steam donkeys were driven out of business by their aggressive competitors. The mills, meanwhile, were modernizing with double and triple circular saws, band saws, edgers, planers, and a hefty assortment of log-moving machinery. The larger outfits were better equipped to manufacture a wide variety of products, just as they were better equipped to extend their railroads and donkeys deeper and deeper into the woods. Indeed, the most successful companies were vertically integrated to include not only the extraction and manufacture of the lumber, but also its transportation and sale in distant markets. The big companies owned the timber, the logging railroads, the mills, the lumber schooners that cruised the Pacific Coast, and the distribution yards in major population centers such as San Francisco.

Back in the woods, the logging crews were larger than ever. Cable logging was a complex affair, requiring precise coordination among the separate tasks. When the choker setters had attached a log to a cable, the whistle punk signaled the donkey puncher that it was time to reel the line in. Once the log was on the landing, the chaser unhooked it; the loaders and railroad men took over from there. For each logging show, there was a variety of specialized personnel: hooktender (woods boss), saw filer, donkey doctor (mechanic), and a full-time bucker just to supply the fuel for the steam donkeys. With the camps now housing a

minimum of 50 men—sometimes up to 200—a host of service personnel was also required: storekeepers, time-punchers, bunkhouse
lackeys, and, of course, the cooks. The companies became wholesale
suppliers of food, clothing, and other amenities. The same railroads that
carried the logs out of the woods returned bearing the necessities of
everyday living.

There were more men on the job—and more machines. With cables
and logs flying through the air, split-second timing and communication
were paramount. Yet speed and production, not the safety of the
workers, were the prime concerns of the highball shows; many a highball logger found himself in the path of a log or a swiftly moving cable.

Death was an ever-present reality for the workers in the woods. In the
words of Walt Hallowell, who once worked on a record-breaking crew
which yarded and loaded 1,690,000 feet of timber in an eight-hour day:
"I've heard of a man getting killed in the morning and they'd put him
behind a stump until that evening to go into camp. Now that's what
they call a highball camp."[7]

It was no accident that the increased influence of the IWW and other
radical union organizations in the woods coincided with the climax of
the highball shows. The workers were subordinated to the dictates of
productive efficiency. The individual lumberjack felt—and was—of little
importance. All large shows had several "utility men" waiting in the
wings to fill in for a sick or injured worker, because the interdependent
tasks required that each and every position be filled at all times. Jess Larison recalls: "Several times I've taken a man's place where they've just
been killed on that job, and stepped right in and did his job. It don't
make you feel too goddamn healthy or secure."[8]

Of all the jobs involved in donkey logging, none was so dangerous—
or so dramatic—as high-climbing. Once a spar tree had been selected to
support the elaborate network of cables, a man had to ascend it with
spurs and a rope to cut off the top. It took an adventurous spirit to work
in the woods in those days, and high-climbers tended to be the most
adventurous of all.

Ernest Rohl

High-Climber, Donkey Puncher, and Timber Faller

"I came by way of Cape Horn. Cape Horn to Callao, Peru, and
from Peru we went up to Tacoma to take a load of lumber to go
back. But I jumped ship there. I wanted to see another country.

"I started to sea when I was fifteen years old. Had my first drink of whiskey when I was fifteen, going around Cape Horn. The cook can't cook, the crew got to have something, and everybody got a tin cup full of rum. And hardtack, ship biscuits. The cook can't cook because as soon as he gets a fire started, a wave comes over, the fire's out. But you take a fifteen-year-old boy with a tin cup full of rum, and he don't give a damn if the ship go down and never come back up. That was 1908.

"In 1915 I came to Eureka from Callao. I was second mate. Then the immigration commissioner advised me to quit the sea, because I was not an American citizen. So I was looking for work. Work's hard to get. I'm a sailor, right? What the hell does a sailor know on a dairy ranch? So I walked. Walked through Fortuna, and I come to Scotia. Up to the office, a guy says to go to the mill and ask the foreman. I didn't know the mill from nothing, but I went into this building. It was a big building. And pretty soon somebody turned all hell loose in there—all the damnedest racket! Must have been right after lunchtime, I don't know. All the machinery, everything, was squealing. Somebody come in and asked me what I was looking for. I told him, 'I'm looking for the place to get the hell out of here. Nothing doing!'

"I took out of there, up the county road to Holmes Flat. There was a new camp up there. Section foreman asked me if I had cork shoes. I could talk good English, but 'cork' and 'calk' shoes I didn't know. 'Oh,' he says, 'never mind.' Takes me to this tall fellow over there, says, 'Put the boy to work, he needs work.' So I got a job bucking wood on a pile driver.

"Then came first rain in the fall: that's it, no more, the woods shut down. So I came back to Eureka. I ran across an Englishman I sailed with on one of the British ships. I said, 'What are you doing, Bill?' He said he was 'on the beach.' When a sailor is 'on the beach,' he's broke. He doesn't say, 'I'm on the bum.' He's on the beach, like driftwood. So I paid his bill—owed a month board and room—and we had a drink of beer. Next day, we walked up the street and he said, 'Ernie, there's a schooner looking for a crew. Bound for Sydney, Australia.' I said, 'Bill, are you all right? A limey and a sauerkraut—going to Australia?' That was World War I. So he says, 'Oh, shank your bloody nationality.' So I shanked my nationality from a German to a Hollander and I signed on.

"Next year I landed back in Eureka. I ran across the pile driver boss and went back to the Pacific Lumber Company. Worked on the pile driver there for a couple of months, then that was all the driving. Had all the trestles done. I got on the spool donkey, pulled rigging. Little side-spooler. Single-engine, just one engine on one side. Pulled the logs in with a cable. You pulled the line out on your back, and the donkey pulled it back in with the log. Used eleven-sixteenths-inch line, eight strand.

"On the nineteenth of June, I had an accident—another young fellow and I. Rigging let go of the stump. They didn't use straps, they used grabs, chain grabs. There was a hook on one end, then about three foot of chain. Put that hook in the stump, then you hook your pulley in the end of the block, and then your rigging. Well, one of them chains pulled out of the stump and let go and killed the guy standing alongside of me. Cable knocked me forty feet, killed the other guy. Young fellow from Arcata. The only thing that killed that fellow, he had a habit of either sitting down or leaning up against something. Him and I was side by side, only I was standing free and he was leaning up against a windfall stump, and he took the solid blow. There was no give, he got the full force. I was loose.

"Well, I lost couple, three days, then went back on the rigging. You put your yarder ahead. They got them eleven by thirteen donkeys, some places got bigger donkeys. They pulled those logs to the skid road. The bull donkey is down on the landing. He pulls it down the skid road to the landing. On the landing, load them onto the train.

"The bull donkey goes on the length of his line, 2,400 feet. All right, the yarder is ahead of that, logging to the skid road. Then the water-slinger, he goes up with the bull donkey, ties onto his logs. He builds his own load. Head log got to be a big butt log. They have all the way up to twenty, twenty-eight logs to the load—that depends on the size of the logs. Then that bull donkey will take those logs down to the landing, pull them in. Then he goes back after another one. When that section is loaded, then the yarder moves ahead and the skid road is built ahead another couple thousand feet.

"Old days when they had skid roads, there was no such thing as take your donkey and put it in the car and take it out there and load it. You had to move your donkey overland. Sometimes it took you two, three days to get you from one setting to another. Moved by its own power. You strung out the line, using blocks and pulleys; it pulled itself up.

"Nineteen eighteen the Pacific Lumber Company started high-lead logging. They had a fellow there to raise the first pole. All the guy lines had to tighten the same time—that's how it worked. First fellow broke the guy line and down he come. Then the camp boss tried to raise that spar tree. He was in a hell of a hurry, and he broke it. Then the superintendent said, 'Let that young sailor raise the pole.' I said all right, so I went over and raised the first spar tree—it wasn't only but about 120 feet. We got that rig up, put the block up, and the mainline. Pretty soon, they broke a mainline. Look who gets up—that sailor. The sailor had to go back up and put the lines through. Every time they broke the lines, sailor had to come.

"I was boss on one of them little spool donkeys. So one day I asked the superintendent, 'Say, how's the chance to get fifty cents more a day?'—because I had to go up the pole every time the line broke. He said, 'No, if I raised your salary the board of directors would fire me.' The next morning the little sailor was gone. I went up in the hills. Stayed there for a week. Come back down, the superintendent wants to see me. I started back for the hills, and he come out from behind the stump, him and the head timekeeper from the main office. 'Ernie,' he says, 'we raised your wages from four and a half a day to two hundred and twenty-five dollars a month. But we don't want you to quit every time a fellow drops his hat.' I say, 'You tell them guys not to drop their hat, and I won't quit.'

"Then I start topping the trees. Redwood. They would like to have the high-lead block 200 feet off the ground. So you always had to top your trees at 210, 220. You shimmy up, top your tree, put your block up, and put up your rigging. That's your spar tree on the landing. Felt just like the mast out on the old sailing ship at midnight. You don't see no landing, no nothing. I could go up the tree fine.

"Woods accidents, they used to have some bad ones in the early days. Lots of fellows got their legs broke, pushed between logs. I had this leg broke in '28 and busted the other one in '29. In the old days, a lot of the accidents was cables. Either the rigging pulled out of the stump, or the strap broke, or something. Or logs. Some got the logs rolled over, pinched in the logs. Quite a few got killed with widow-makers, limbs. Sometimes a loose branch hangs up there, and at the least little movement down it comes and hits you.

"One time we were working on a family tree. We had a saw in there, and pretty soon it stuck. We had to beat it off by hand, chop

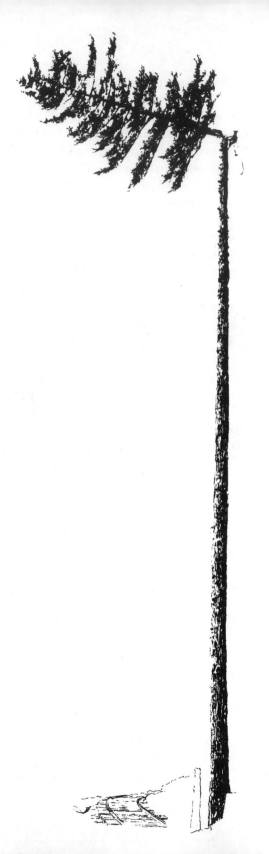

Raise the pole: high-climbing

with an ax. So Ernie [Ernest's son] started in, and I started in on the undercut. Pretty soon that thing starts wiggling. 'Ernie,' I say, 'you get the hell out of here.' It was a family tree, a group together, all tangled up. He says, 'You don't know what's coming.' But I say, 'By God, I do.' I gave it two more licks and the tree popped. I took off. Ernie hollered, and it's a good thing he did: 'Dad, look up!' I just froze right there and I looked up and a chunk hit me in the kisser. If he didn't holler, I would have had it right between the ears. Got my eyes, bashed in my nose, everything. Knocked me down on the ground. My hat was gone, and whenever I'm without my hat I'm lost. I say, 'Ernie, Jesus Christ, get me my hat.' The eyeball it was popping around there someplace, popped out. Then I had to walk back up the hill.

"The worst mess I seen was a fellow and his partner had a forked tree. Put in a couple of sticks of powder to blow 'em apart. Tree didn't fall when the shot went off. Fellow started back down toward the tree. All at once it started to pop. Henry went one way, partner went the other way. That would have been fine, but he changed his mind. He doubled back, and just when he came out from behind, it came down on him. That man, if I didn't know him I couldn't have swore it was him. The only thing was his scalp. Flattened him out. One arm lay forty feet from him.

"One railroad accident I know of. Man of German descent, Fritz Speck, a superintendent from the Pacific Lumber Company. They had a derailment. In the back end they had a car with a bunch of hogs in it. Around this old-time camp, they had a cookhouse that fed 200, 250 men. They'd take pigs up there, fatten the pigs. This train had a hog car, and in back of that they had a caboose. So Fritz Speck, he was there in the caboose and broke his leg in the derailment. The brakeman says, 'Mr. Speck, how are you? You hurt? You hurt?' 'Never mind me,' Speck says. 'You watch the hogs. Don't you let *them* get away. Never mind me.' He had a broken leg, but he didn't give a damn about that as long as the hogs stayed there. I admired that fellow.

———— 99 ————

Internal Combustion: Tree Mining Comes of Age

The railroad days in the woods are over. Gasoline power replaced steam power in the second quarter of the twentieth century, making a dramatic impact on both the woods and the men who worked there. The internal combustion engine came in three distinct forms: the chain saw, the logging truck, and the caterpillar tractor. Chain saws replaced the old misery whips, enabling every modern-day logger to outcut even the best of the old-time lumberjacks. Logging trucks replaced railroad cars, eliminating the need to lay down and take up tracks every time a logging show had to switch locations. Caterpillar tractors replaced the old steam donkeys, eliminating the need to construct an elaborate cable network for each neck of the woods. The same caterpillar that yarded the logs could also build the roads. Far more mobile than its steam-powered predecessor, cat/truck logging overcame previous topographical and geographical limitations. The cat/truck combination was the final answer to the problem of accessibility: there is now no area too remote for a viable logging operation. The harvesting of timber can—and does—occur anywhere and everywhere, not just a mile or two from the nearest river or railroad.

Because the trees could now be cut more quickly and in a greater variety of locations, gasoline power hastened the depletion of old-growth timber stands. Timber companies and gyppo loggers accelerated their cutting rates on private land as new roads opened up rugged terrain in the backwoods. Starting in the 1940s, the market had no trouble absorbing the extra production: World War II and the postwar housing boom created an unprecedented demand for lumber. Cutting proceeded apace for about twenty years, but by the 1960s the effect of stepped-up production was becoming apparent: privately owned, old-growth timber was on the verge of extinction.

In order to maintain a supply to meet the demand, loggers had to shift their harvesting operations to public lands, which had been removed from production by the conservation movement in the early 1900s. In the 1930s, less than one-third of the timber harvested in the Pacific Northwest came from public lands; by the early 1960s, public lands accounted for more than half the cut.[9] As the burden of logging in the West shifted to government lands, mill owners and lumberjacks found they no longer could determine the fate of the woods by themselves. Forestry had become a matter of public policy. Although new tools had given the loggers more power to overcome physical obstacles, a combination of dwindling resources and public pressure began to diminish that power.

The increased harvests generated by gasoline power spurred a second wave of interest in forest conservation. When logging trucks and caterpillars started their penetration of the backwoods in the early 1930s, the federal government once again took tentative steps toward regulating forest practices. In 1934 the Lumber Code Authority, established by Franklin Roosevelt's National Recovery Act, called for regional conservation rules that would prevent damage to young trees during logging operations and provide for the replanting of cleared land. A few of these regional regulations were in fact drafted, but they quickly became inoperative when the National Recovery Act was ruled unconstitutional a year later.

Despite this setback, the Roosevelt administration showed a continuing interest in regulating logging activities and, perhaps, even nationalizing privately owned timberland. The timber industry, wary of federal restrictions and afraid of nationalization, began to push for state legislation that would be less threatening and easier to control. Ironically, the industry itself helped draft the early forest practice acts in the West. In 1941 Oregon passed a law requiring loggers to leave 5% of the original stand as seed trees; in the pine country in the eastern part of the state, the law required that all trees less than sixteen inches in diameter remain standing. In 1943 California passed its own minimum diameter law, which prohibited the commercial cutting of coniferous trees less than eighteen inches in diameter. In 1945 Washington passed a law similar to that of Oregon; California, meanwhile, developed a unique Forest Practice Act in which all rules were to be voted on and approved by property owners representing two-thirds or more of the state's timberland. The rules adopted in this democratic manner were not exceptionally forceful, for the landowners showed little interest in restricting their own activities too severely. The California law lacked any enforcement provisions and was administered by a Board of Forestry dominated by industry representatives.

The impact of these regulations was minimal—and in some cases even harmful. The loggers did not mind leaving a small percentage of trees behind; often, they would have left some trees anyway simply because they weren't worth taking. A few conscientious loggers left top-quality trees, but many preferred to take the good ones and leave the bad. This practice of "high-grading," as it is called, amounted to a reverse genetic selection: only the inferior trees were allowed to reproduce. Regeneration from seed trees ran into other problems: by the time a good seed-bearing year came along, unwanted brush and noncommercial hardwoods had often taken over the land. The primitive techniques for regeneration, as reflected in these early forest practice laws, continued to lag behind the advanced techniques for harvesting.

The impact of the accelerated harvest was felt by the loggers as well as the forests. Inevitably, the local booms in the backwoods were followed by busts. Employment ran high for a time, but then the work was gone.

Loading the truck: Oregon, 1942

With cats and chain saws at their command, the productive output of each logger increased, but the number of men employed in each logging show was correspondingly lower.

Many of those lumberjacks who maintained steady work became commuters. With roads to and from the logging shows, men could live at home and drive to their place of work. With no need to live in lumber camps, it was easier for a logger to be a family man. Even if work could only be found far from home, the whole family might pack up and join the logger in a house trailer. For the logger whose place of work must change year by year, trailers and mobile homes have provided a compromise between a transient and a settled way of life.

On another level, the takeover of the woods by machinery has deromanticized logging. To fall and move a ten-ton tree is no longer the ultimate challenge for manly strength. The ancient cry of "timber" cannot now be heard over the motorized hum of chain saws, cats, and trucks. In the words of Bob Ziak, Jr., a lifetime logger along the Columbia River:

> 'Course now, today, we have the power saws, and they make plenty of noise, but falling timber was comparatively quiet. You'd just hear the swish, swish, swish of the saw and it wasn't even that loud, and then once in a while you'd hear the ring of the wedges as they'd wedge their tree. Today I don't hear the sound or the cry of "timber" . . . and it was a beautiful sound.[10]

For a pair of old-time timber fallers working by hand with their misery whip, a giant tree was a formidable adversary worthy of great respect: "He's one of us. He's tree. I mean it's you, the next fellow, and the tree. There's three people, I guess you might say."[11] Today, the trees fall quickly, one after another, leaving a man no time to ponder their demise.

2

TREE FARMING
The Voice of Industry

To meet the accelerated demand for timber created by World War II and the post-war housing boom, the coniferous forests of California and the Pacific Northwest were cut at an unprecedented rate. The imminent depletion of old-growth timber presented the industry with a new challenge: how to re-create a resource base for the future. If the companies wanted to remain in business, they would have to shift their focus from the harvesting of ancient giants to the nurturing of tiny seedlings.

Of necessity, timber interests began to develop a new perspective. The few virgin forests that remained were valued for more than just their standing wood fiber; they were seen as productive sites that could support future crops of trees. Forested land was appraised for its potential, considering the trees that might be grown as well as the trees that were already there. The old "mining" approach to timber had been static: once the existing resource had been depleted, the forest was worthless to the loggers. Suddenly, the concept of a forest became more dynamic: growth rates were as important as standing inventories. The emphasis switched from extraction to production.

Old-growth trees were still cut down, but the logging was supported by a new rationale. The timber industry, in its concern for production, regarded virgin forests as wasteful: the rate of decay in "overmature" timber often equaled or surpassed the rate of growth, bringing net productivity to a standstill. In the absence of logging, the rotting trees would be harvested by natural means, such as wind, fire, insects, or disease. Logging operations, therefore, served to salvage dying timber for human needs and reclaim the land for productive purposes. The old growth was liquidated to make room for young, fast-growing trees.

The mammoth timber from an untamed, unproductive virgin forest ceased to be the logger's dream; instead, he came to prefer the rapid growth and high yields of a scientifically managed tree farm. In the words of a 1947 forestry textbook, "Timber resources, untouched by man, present a striking parallel with wild prairies awaiting the plow."[1] Just as wild grasses were once removed to open up the land for domesticated crops, so did nature's random assortment of trees have to be cut down to facilitate the farming of timber. The comparison of forestry with farming gave rise to a new terminology: *plantations* of *crop trees* were protected from insects, disease, and *weed trees;* eventually, they would be *harvested* in cyclical rotations.

If forestry was the equivalent of farming, the timber industry reasoned that it made no sense to set aside areas where the tree crops were not allowed to be harvested. Commercial foresters came to see the preservation of forests in their natural state as a waste of productive timberland. An advertisement by a timber company in a 1977 forestry journal showed a picture of a single ear of corn with a $3.25 price tag:

> Suppose farmers found corn too lovely to pick? Or imagine the rancher so moved by the sight of his wheat field that he canceled the harvest.
>
> Sooner or later, all crops are harvested. If man doesn't use the bounty for himself, Mother Nature will claim it. Via insects, disease, rain, high winds or low temperatures.
>
> It's easier to understand when we talk about food crops. Because we can witness the outcome within a single year. With timber the cycle is a lot longer. But the laws of nature are the same.
>
> So are the laws of economics. Because whenever large portions of a crop aren't harvested, the price of that crop is going to rise. It's true for corn, apples, peas, and beans, just as it's true for the timber that produces lumber, plywood, particleboard and paper products . . .
>
> To be sure, the forest holds enormous beauty. But it also holds great promise. To realize that promise, we must remember that there is a time to sow and a time to reap.[2]

The same theme was expressed more directly by H. D. Bennett, a timber industry executive from the Appalachian region: "We have the directive from God: 'Have dominion over the earth, replenish it, and subdue it.' God has not given us these resources so we can merely watch their ecological changes occur."[3]

The idea of tree farming was not that new. Forests had been farmed in Europe for centuries, for the virgin forests there were depleted long ago. In this country Gifford Pinchot, the famous conservationist of the

The tree farm: rank and file

early 1900s, was fond of using the agricultural model: "Forestry is Tree Farming. . . . The purpose of Forestry, then, is to make the forest produce the largest possible amount of whatever crop or service will be most useful, and keep on producing it for generation after generation of men and trees."[4]

Yet the idea of tree farming did not really take hold until the old-growth timber in the West showed signs of imminent exhaustion. In the early 1940s the price of lumber soared, giving timber interests plenty of available capital—and a sufficient incentive—to invest in reforestation. No longer were there untouched parcels of old-growth timberland that could be purchased cheaply on the open market with the profits reaped from the depleted forests. The only real alternative was for the companies to manage the available private land to produce future crops. Thus was the American Tree Farm System born.

Bill Hagenstein

Executive Vice-President, Industrial Forestry Association

"When I got out of graduate school, I got an offer from the West Coast Lumberman's Association to be their forester in western

Washington. I went to work on the sixth day of June, 1941, and six days later Weyerhaeuser announced that they were going to manage a property in Grays Harbor County as a tree farm. Our association picked up the idea that Weyerhaeuser developed there and made a program out of it. In the fall of 1941 we recommended to the National Association that they do something similar for the rest of the country. On the twentieth day of January, 1942, our board of directors certified the first tree farm in the United States. I was in the room when it happened. It was right up the street here in an old building that was then called the Portland Hotel. It's a parking lot now, but I'm going to get a plaque put up there someday in the right place, because there's where the American Tree Farm System started. And I was there when it did.

"Up to that point, there had been many attempts in the United States to practice forestry. Most of them were not very successful for a number of reasons, one of which was that we still had hopelessly inadequate protection against fire, so it was impossible to get the government or private owners to invest money in growing trees on purpose, because there was no assurance that when harvest day rolled around there would be anything to harvest. In the meantime, they may have burned up. One of the pioneer companies in this region, the Long-Bell Lumber Company of Kansas City, came out here in 1920 with the idea that they clearcut this old-growth timber and replant every acre, and they started to do it. They built a nursery and when they started operating in about 1923, they planted the first trees, and every year thereafter they planted the areas they cut. But the one thing they couldn't do— they were ahead of their time—was guarantee that they could prevent those plantations from burning up. And in 1938 the whole goddamn bunch of them did. They had a bad fire in there and burned them all up. It discouraged the hell out of that company from doing it for a long time. It took almost fifteen years to get them interested in it again, because they felt they'd be throwing good money after bad.

"When we certified those first tree farms, the whole idea was to get strong public support for improving our protection against fire, for recognizing that it takes a long time to grow trees. From the time you plant a little two-year-old tree to the time it's ready to harvest, every year you run the uninsurable risk that something could happen to it: it could burn up, it could blow over, it could be killed by insects, it could be stolen by somebody if it's accessible. And you've got to pay taxes every year on the land. Taxes, protection costs, and the accumulation of risk year after year don't add to the value of timber—they only add to its cost. So all during this period, everybody was speculating.

"Those of us hired by industry used the tree farm program as a vehicle to get public support for good protection, for reasonable taxation. At that time there was a drive on in the United States—a political drive by the Roosevelt Administration—to allege that a long-term crop like timber couldn't be handled by anybody except the government. The government would either have to grow timber on its own lands alone or would have to regulate the private owners. And there was nobody in our industry who looked with favor upon the idea of the federal government coming in and telling us how to do it. So the tree farm program was in part a vehicle to build up some public confidence that here was an industry prepared to do the job of managing these lands.

"My job, and the job of our association all the way through this thing, has been to encourage the people to grow trees, to stimulate their interest by showing them what their opportunities are, and then suggesting to them what they have to do to realize those opportunities. And the record really speaks for itself. For example, in our region we have certified as tree farms more than 60% of all the privately owned forest land in western Oregon and western Washington. To become certified, we require any private landowner—and it could be you or me—to agree to keep the land for the purpose of growing commercial crops of timber. You agree to provide it adequate protection against fire, insects, and disease, or damage by destructive grazing. And you also agree, when you harvest timber, that you harvest it in such a way that you keep the land productive. What that means is that you're going to get it reproduced properly, either artificially or naturally. You can do it either way, but most people today reproduce artificially.

"Originally, a lot of our plans thirty years ago depended heavily on natural reforestation; you cut the timber in such a way that you left the seed sources. But it soon became clear that we couldn't run the risk of letting that land be occupied by brush species. The Old Man Upstairs had built the land into a coniferous forest area; we ought to keep it that way, and not let the hardwoods get it. There's a cycle to the seed crops on the conifers that only averages about two reasonable crops in a decade. If you cut the timber in a period when you have two failures and two light seed crops in a row, you're not apt to get adequate natural reproduction. You get some, but you won't get enough. You end up with a partially stocked stand of timber, and you don't succeed in forestry by having the shelves only partially filled. You've got to fill them up. Better to have too many trees than too few. It's just like your own garden: you want to plant them thick, and if you get too many, you're going to thin them out. You don't want to waste space, otherwise you don't get much to eat. Same way with trees.

"So at the same time we started the tree farm program in 1941, the West Coast Lumberman's Association started a tree nursery. Up to that time, there was no source of trees for the industry. There were a couple of companies like Weyerhaeuser and Long-Bell that had nurseries of their own, and the federal nurseries were not allowed to sell trees to private owners. So we started a nursery in which we got half a dozen companies to agree to buy trees at a predetermined price, and we announced to the world, bragged to the world, that we were going to grow five million trees a year for reforestation. There was no question that part of the idea of getting the industry to put up the money was public relations, but we didn't let it degenerate into that alone. We turned it into a serious program to really get these unstocked lands reforested. And, of course, it's grown like mad. We started out with the desire to grow five million trees a year, and now we grow forty to forty-five million. Being a nonprofit corporation, we sell the trees back to the companies at cost.

"When a tree farmer gets certified, he gets a paper signed by the president of the association, and then he has the right to call on us anytime, free of charge, for any professional assistance that we can give him: reforestation, protection, utilization, anything to do with the management of that property. Mostly nonindustrial owners are certified as tree farms. We spend a lot of money on some of them, but the industry is willing to have us do that, because, to the extent that the landowners outside the industry are doing their job in keeping their land productive, this is going to add to the sum total of timber supply, which ultimately the industry is going to have a chance to buy. That's important to us, because in this industry half of the manufacturing entities in it don't own an acre of land or a stick of timber. They're entirely dependent upon public timber or timber from other private sources. It's only the landowning companies that are anywhere near self-sufficient.

"For years, the big argument for many people *not* to invest money in forestry—particularly for an individual—was that you'd never live to see the results. It took too long. Today if I have twenty, forty, or sixty acres of young trees coming along, and I decide that I need that money to do something—pay medical bills, send a child to college, build a house, anything that requires some capital—why, I'd find a ready market for that timberland. The trees don't have to be mature and ready for harvest. That's changed the picture, that's encouraged a lot of small individuals to go into tree farming. You'd be surprised at the number of businessmen that come to see us each year wanting to go out and get a piece of land and grow trees. I've got a piece myself, a twenty-acre piece up in Washington, and I can say it's the best investment I ever made in my life.

———— 〃 ————

Taming the Forest: Even-Age Monoculture

The key to tree farming is intensive management. The complex forest environment is controlled and manipulated to maximize the major element of commercial interest: wood fiber. As Bill Hagenstein puts it, "Forestry has always been an environmental undertaking. Its main thrust has always been taming the wild forest for man's use and enjoyment by managing the ecology, instead of letting it run rampant as though there were no people around."[5] The tools and techniques of modern agricultural science are brought to bear on the woods, tending and caring for the well-being of the crop trees. In the words of Bernard Orell, former Vice-President of Weyerhaeuser, "This means the application of fertilizers, insecticides, and herbicides, much as we nurture children to the full flower of adulthood by use of medicines, nutrients, and preventatives."[6]

The most significant improvement that man can make over nature is to shorten the rotation age for each crop of trees. Nature provides for regeneration, but it does so only slowly. After a fire, windstorm, insect infestation, or other natural means of harvesting the trees, the forest is opened up for fast-growing, sun-loving species of plants. These *pioneers*, as they are called, penetrate and enrich the soil, shade the exposed earth, and start to build back the forest biomass by photosynthesizing the sun's energy. Pioneer plants are relatively short-lived and often of limited size; during their period of dominance, however, they serve to create an appropriate environment for the various successor species. Several stages of plant growth may ensue, culminating in a climax forest of mature trees. This process of forest succession sometimes takes centuries to work itself out.

When a mature forest is logged, nature's plan for forest succession is set into action. Timber growers, however, can accelerate this lengthy process with various management techniques. If the only trees with major commercial value come late in the cycle, they can try to eliminate the pioneer stages through some form of weed control. They plant the desired trees, even if they are successor or climax species, and they do what they can to eradicate all competitors. By imposing a one-step cycle in place of the complex process of forest succession, forest managers can reduce the time required for each crop to half a century, or even less. In the words of Dave Burwell from the Rosboro Lumber Company, "I know it took nature five hundred years to grow that forest, [but] we can do it in fifty, because in this climate the damn stuff grows back faster than we can cut it."[7]

While shortening the rotation, tree farmers are also simplifying the ecosystem so it is easier to manipulate. Rather than waste space with noncommercial species, they like to stock the land with only the most desirable trees. And the behavior of the trees is more predictable—and therefore more easily controlled—if they are all of equal age. With even-age trees of the same species, the techniques of intensive management can achieve their most direct results: fertilization and weed control can be applied only when the trees are most in need of assistance, and the trees can be treated for specific diseases at the most susceptible stages in their development.

This sort of even-age monoculture is modeled after standard farming practices. A cornfield is intended to raise corn—and nothing else. The farmer does not plant an occasional pea or bean amidst the rows of corn, nor does he permit the field to be overrun with dandelions. He plants his seeds all at one time so his crop will mature evenly. Come fall, the farmer does not harvest every fifth cornstalk and leave the rest; he systematically "clearcuts" his field.

And so it is with trees, assert some members of the timber industry. Clearcutting is an integral part of even-age monoculture. The surest way to create an even-age stand of a single species is to remove the old forest in its entirety and start with a clean slate. All trees must be cut to the ground, whether or not they are commercially useful. Then the residue must be cleared away in order that the soil be exposed for planting. The industrial forester, like the farmer, prefers to work with bare earth. Corn seeds are not sown amidst weeds and last year's stalks, nor are young trees planted in brush and logging debris. Industrial foresters use a full arsenal of tools to prepare the site for planting: they use heavy equipment to pile slash and scrape the soil; they use fire to burn the slash and gain easy access to the ground; they use chemicals to kill weeds and remove pathogens from the soil.

Once the forest cover is totally cleared, tree farmers are ready to plant their next crop. The new trees, however, will not be like the old ones: they will be faster-growing "super-trees," the genetically engineered products of seed orchards and scientifically managed nurseries.

Philip F. Hahn

Manager of Forestry Research, Georgia-Pacific Corporation

"Genetics is an old technique used in agriculture and animal breeding for centuries. In crops like wheat and corn you get a crop every year, so you can make rapid advances in genetics. If you have a selection program, a crossing program, you can see your results a year later. Animal breeding is a little slower, but even with animals it takes only a few years for a generation. But when it comes to trees, you have to wait for decades before you see actual results from a genetics program. That is probably the reason why genetics in forestry wasn't applied as rapidly, and still isn't applied as easily, as in other areas. But companies and foresters realize that we must get into this field if we want to think of the future.

"In the fifties, Georgia-Pacific and some of the other companies started on these new genetic programs, but we could not produce enough seed in a genetics program to spread the seed from a helicopter. We had to find a new way to reforest our land, and that new way is producing seedlings from a genetically improved source in nurseries. With nursery stock, we can spread the seed over a larger area. In aerial seeding, we use roughly twenty to forty thousand seeds per acre and get maybe a thousand trees, which normally would be poorly distributed over the area. That's wasting seed— and land. But with hand planting, we can spread the same amount of seed over thirty or forty acres quite evenly to cover all usable area.

"Let me explain how a genetics program works in a practical application. Trees, due to natural development, are adapted to local conditions. This is recognized by foresters, so we use the natural stands in a given area because we know that's the best stock there is. The trees survived there for centuries; if they evolved there naturally, we really cannot do better. Of course, there is still a lot of variation among trees within those stands; you still see some runts, some average trees, and some outstanding trees. In a genetics program, we are after those very outstanding trees. We choose the very best stands in each local area, and within the very best stands we choose the very best groups of trees, where there is a lot of competition, and out of all that competition we choose the very best individual tree. We call this tree a 'plus' tree.

"We could go out there to these trees and collect cones and seeds, but they are spread all over the area, and they are generally tall, mature trees. We would have to climb up and get the cones each time we had a cone crop. But we don't want to do it that way. Technology provides us with a tool by which we can reproduce those trees identically through a vegetative propagation method. We have to get a part of that tree, and the best part is the top of the tree, because that represents the best cone-producing ability. We go out there and take the top off and take cuttings off the tips of the branches. Those cuttings are then used to reproduce the 'plus' tree through grafting.

"Before we develop our cone-producing orchard, we develop root stock that will be compatible with the grafting material. With a compatible root stock, we are able to reproduce these trees easily and successfully. The grafting helps to bring the trees into almost instant cone production. The cutting that was grafted onto the root stock still thinks it's about sixty years old, and it goes on producing cones even when the grafted tree is only two or three feet tall. In an orchard, this is very important because we try to collect cones while we stand on the ground. We don't like to climb trees if we don't have to. We prune them back, but sooner or later the trees will get away from us because they're fast-growing trees, and then, of course, we'll bring in machinery to help in collecting the cones.

"The orchards are generally located on relatively flat terrain. We can easily plant the trees out there, we can thin the trees, we can irrigate them, we can fertilize them, we can cultivate around the trees, we can use mechanized equipment in taking care of them and for collecting cones.

"There is another advantage to having an orchard. Imagine: you have 300 outstanding parent trees, and they freely mix with each other. They improve upon themselves through natural crossing. We also use controlled pollination. This way, we are able to test and further select to improve even on the best trees. Naturally, we have to test them out carefully. They are all good, but we want to find out which are the very best. The testing is done on an ideal site, which we call a progeny test area. These tests give us a lot of information about the trees. We are interested not merely in how they appear in the forest at selection time, but how they pass on their traits to their offspring.

"We want to find out which trees grow rapidly in the early stages, because Douglas-fir needs the sunshine and the trees need to grow rapidly, so they can compete with the brush by staying on top. We test for actual growth rates, height growth, diameter growth. And, of course, we are interested in trees that are disease-resistant. They might enable us to avoid using pesticides. Disease resistance is easy to test for: we inoculate the trees and subject them to disease, and the ones that are not receptive to the disease will survive. We also look for desirable branch characteristics, and we test for specific gravity, the density of the wood. Trees with higher specific gravity will provide a stronger wood and will produce more pulp, so we'll have a higher return. We can test for all these things. It's all done scientifically in a statistically sound system. This way we can sort out the trees that have good wood quality, are disease-resistant, and grow rapidly. Of course, nature always has done a good job of selection, but we expedite the selection ourselves. Foresters are keen observers: they learn from nature and they soon figure ways to manage the forest better than nature does.

"In order to do a large reforestation job, we have to have a proper nursery facility. In the past, standard practice was to produce bare-root seedlings, which were grown out in an open field for a couple of years and, when they were ready, were lifted, packaged, put in a bag, and planted. That was a well-developed system, but we are in a new era now, and technology improves, and we soon found that there are other ways to produce seedlings in large quantities. So we got involved in containerization.

"The container technology was still in its infancy when we started developing our nursery facilities in 1970. There was no equipment available, there was no proven greenhouse system. I knew the trees needed some artificial help, but I knew I didn't want to baby them. They have to go out in the field, and they have to stand up under very adverse conditions. So I designed a greenhouse system that works like a convertible car: when it's cold, you close it; when it's warm or hot, you can open the sides and the roof. We even put a shade screen over it, so the trees will feel more comfortable, but will still be exposed to a breeze moving through the greenhouse to keep the disease problem down.

"To produce several million seedlings sowing by hand would take a long time. So we had to design equipment to do this. We invented the seeding equipment, so we can set up an assembly line to move the blocks of containers through rapidly. With our sowing system, we are able to seed about 300,000 cavities a day while using ten people. This is considered a fairly good production rate. This way we can get the job done in about six weeks, even if we have to sow about ten million trees, as we do right now.

"When we started out, we were skeptical about how to plant out in the field. We went back to the drawing board. I designed a backpack to carry the blocks (styrofoam containers); this pack can eject just one quarter-section of the block at a time. Then a tree planter can place the quarterblock in a belt holder in front of him and have direct access to the trees. He can just pull on the seedling, and the whole plug of dirt will easily slide out, because the container cavity is tapered. With a dibble he can make an impression in the ground the same form and size as the container plug, and then just place the seedling plug in the soil cavity on the cut-over land without disturbing the root system. The tree often does not know the difference, whether it is sitting on a nursery bench or is out in the field.

"There are a lot of angles to containerization. In an outdoor nursery germination takes a long time, some seeds germinate rapidly and others don't, and you get an uneven crop. In a container nursery, they all germinate fast because of the artificial help, and you get an even crop. They're off to a good start, and you can gear the watering program and the fertilization program to evenly germinated and evenly spaced trees. Since they have equal growing space, they can grow rapidly with no interference

Test-tube timber: a
Douglas-fir tree plug

from the neighboring trees. If we sow in the beginning of April, we won't have any difficulty in reaching the maximum height by the middle of July. Then we taper off with our growing and start our hardening program. These trees have to go through a hardening phase, because they have to be conditioned to go to the field. This way we can have them ready for planting in the winter, when the trees are dormant. We can grow one crop in one season, and that's important so we can utilize our nursery facilities well and are able to schedule our reforestation better. We don't have to wait two or three years to get a crop.

"Before we actually plant the seedlings in the forest, we have to clear the site. A successful reforestation effort actually starts with the logging. After harvest, the limbs and the tops of the trees and some of the other brush stay behind in the form of slash. This can cause a problem in getting onto the land physically. So, it's common practice to burn some of this material to make the area accessible to the planters. This also exposes the ground. Douglas-fir likes an exposed soil. This is known through experience; the companies are not just burning for the sake of burning. We have to prepare the site, and, of course, nature often prepared the site by burning, which resulted in outstanding crops.

"Often, you have too many brush species in the forest. In a case like this, you have to go in and use mechanized equipment to actually

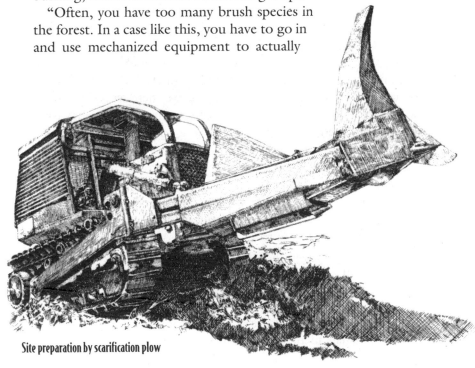

Site preparation by scarification plow

remove this material—either cut the stems down, or push them down, pile them, and get rid of them. We really don't like to do this if we can avoid it, because it's expensive; but, on the other hand, it does open up the ground for cultivation work. It's just like a farmer who plows his field. He's not destroying it, he's just loosening up the ground and preparing it for a seedbed, and then he plants it. You aerate the soil to get a better start for the new forest.

"Spraying is another important part of reforestation work. Douglas-fir likes a lot of sunlight. If for some reason the other vegetation has a jump on the trees—like in our coastal area especially, we have salmonberry and other fast-growing species—those weed species will get ahead of our Douglas-fir. We have to come in and knock that type of vegetation back to give the Douglas-fir a chance. This is what we call our "release spray": it releases the growth of the tree, which is able to push through the brush vegetation to reach for the sunlight. A few years later, we will get back the ground vegetation, but we have saved the trees, because by then they are above the brush. We remove an obstacle temporarily.

"After the trees are through the process of competing with the vegetation, when they're maybe twenty or twenty-five feet tall, they start to compete with each other. Of course, we like to start with a relatively large number of trees, because this will give us an opportunity to do further selection. By age twelve or fifteen, the trees have shown their dominance, and we are able to go out there and do what we call a precommercial thinning. We send a crew through and eliminate all those trees that would compete with our so-called crop trees, the dominant trees. While we are there with the power saw, we are able to cut down some of the unwanted species too, some of the weed species that are taking up growing space and using the moisture and nutrients in the ground. By spacing the trees, we open up the stands, and the trees are able to accelerate their natural growth rate.

"In order to expedite the growth rate even more, we are moving in with aerial fertilization. Aerial fertilization is used only in areas where we know we have nutrient deficiencies. For optimum growth you have to have a balanced nutrition program. If just one element is missing, you can have all kinds of other nutrients available, and the trees still won't be able to utilize them. By adding the missing nutrient, we put back a balance, and all the other nutrients become available. Of course, fertilization costs money like everything else. In forest management you don't want to waste your money.

There's no point in fertilizing the soil if you don't have to. A company is out to make a profit, there's no question about that. It's a
corporation and it has stockholders. America is a capitalist country,
and everybody wants to make a profit. If we don't make a profit, we
won't be able to sustain a healthy economy. In our organization,
every project has to stand on its own. It has to make a profit and has
to be ecologically sound. If it cannot make a profit, or isn't ecologically sound, we have to look at it carefully for possible elimination.
All management procedures should be as efficient as possible, and
they can only be efficient when we know what we are doing. That's
why, for example, we're testing our soils to find out what fertilizers
we need. And that's why we run all the tests in our genetics program. We try not to leave any stone unturned.

———— 〞 ————

Mothering Young Growth: Protecting the Babies

The tools of industrial forest management can play a vital role in helping
young trees to become established. When the seedlings are first trans-
planted from the nursery, they are easy prey for deer and other browsers
who relish the foot-high foliage. Since animal damage costs the timber
industry in the Pacific Northwest several million dollars a year, the companies like to protect the young trees with plastic tubes or other fencing
devices. Fencing is costly, but a test conducted by Georgia-Pacific
revealed that the growth rates per acre in unprotected reforestation sites
were only 40% of the rates found in areas where all the trees were caged.
By enclosing the seedlings, mortality rates were cut in half.[8] To the
industrial forester, fencing is often worth the extra expense.

An alternative strategy is to coat the young trees with chemicals such
as thiram or BGR (Big Game Repellent), which render the tender
shoots unsavory to animal palates. Theoretically, the animals are supposed to develop a distaste for the treated trees before they become ill
from the toxic effects of the chemicals.

The most direct method of animal control is killing off the foragers.
Gophers, mountain beavers, and other small mammals can be trapped
or poisoned by setting out toxic baits such as strychnine. Forest managers can decrease deer and elk populations by encouraging hunting.
Animals can also be controlled by using herbicides to eliminate their
habitats: pocket gopher populations decline when grasses disappear;
mountain beavers tend to leave an area when swordfern, a major winter

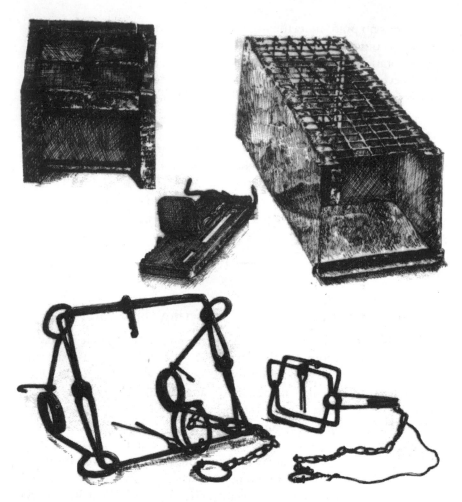

Protection for seedlings: an arsenal of animal traps

food, is removed; all sorts of small, foraging animals find it harder to hide from predators when their brush shelter is eradicated.

Animal damage is only one of the many dangers faced by the young seedlings in their struggle for survival. Like any other member of the forest community, the trees from the nursery are subject to disease and insect infestations. They also must beat out their competitors for the available sunlight, moisture, and soil nutrients. Again, the technology of modern forestry can give the chosen seedlings a boost: they can be treated for disease; they can be sprayed with insecticides; they can receive help in their battle for survival by removing the competition. For every crisis that planted trees may encounter, there is a chemical substance to aid and comfort them.

The most controversial chemicals in forest management today are herbicides, which destroy or retard noncommercial hardwoods while leaving the more valuable conifers relatively unharmed. Applied before planting, they eliminate the food and shelter for undesirable animals; applied after planting, they control brush competition. They are sometimes even used to "rehabilitate" land that has been taken over by hardwoods. In many ways, herbicides are a forester's dream: they involve no soil disturbance, they provide a gradual transition from shade to sunlight as the affected foliage dies off, while they leave standing the scraggly, dead brush as protection against browsers.

In the 1970s and early 1980s, herbicides were applied by air throughout many of the commercial forests of California and the Pacific Northwest. The most common chemical was a combination of 2,4-D and 2,4,5-T—known as "Agent Orange" when used to defoliate the forests of Vietnam. The phenoxy herbicides, as they are called, triggered heated political battles, with environmentalists claiming that they presented serious health hazards and timber interests claiming that they were safe. Today, 2,4,5-T is banned from use, although other chemicals—often applied on the ground rather than from the air—are used in its place. (The current status of the herbicide controversy is discussed in Chapter 4.) The case for the use of herbicides in the practice of intensive forestry was made forcefully by a timber owner from Humboldt County, California, in 1980:

Robert Barnum

Timber Owner

"We have a unique situation here in northwestern California. To understand why we need to use phenoxy herbicides, you have to know the historical background.

"When the early logging was done in California, it was done mostly in the redwoods. But when they started logging fir for the great boom in the housing market after 1945, they came into the Douglas-fir stands, and among those stands were great quantities of hardwoods. They logged by tractor because it was more economical, and they had to leave seed trees because of the new forest practices laws. They went in and cut down the best trees, the big tall, beautiful, straight-grained fir with no visible defects. The trees they left for seed trees were the diseased, the conky, the defective. From a genetic point of view, they were doing the worst possible thing: instead of leaving the superior trees as the genetic parents for the subsequent crop, they left the inferior trees. Also, because of the lack of a market, they left the hardwoods.

"Consequently, we are left today with vast tracts of cut-over timberland in Humboldt, Del Norte, and Mendocino Counties that were logged that way. There must be several million acres like that all up and down the North Coast and into southern Oregon, and on over into Trinity County and Siskiyou. The hardwood trees that happen to be knocked down sprout like redwoods from the stump, whereas a fir has to come up from seed. So the firs were a long time coming up and were from inferior seed stock. But the tan oak trees and all the various species of hardwoods we have here flourished, and they seeded, and they sprouted, and in the meantime the firs were underneath all this. You have hardwoods that exist because of the alteration of that stand by man, not by nature.

"It's an unnatural situation. The people who say they don't want to use phenoxy herbicides because they don't want to disturb the balance of nature—well, they're just about thirty-five or forty years too late. Now herbicides are used to redress an imbalance.

"They're doing things in forestry now that are really exciting if you're into the business or if you live in a forested country. Up in Oregon and Washington, they're growing timber stands much greater than what grows naturally. It's just like wheat. Back when wheat was discovered four or five thousand years ago, and the guy was picking the ground with his stick putting in little seeds, they didn't grow very much for him. Now, you see pictures in the Midwest where they grow it, and they have these tremendous fields, exports all over the world, and so forth. That's because of the techniques they've developed with hybrids, fertilizers, site preparation, and, of course, phenoxy herbicides. I understand that every acre of wheat in Kansas is treated with phenoxy herbicides.

"The same thing is going on in forestry right now. It's intensive forestry. With the price of stumpage having gone up so much, it's economically possible to do these things, where before it wasn't. Now with these new techniques that have been established, practiced, and experimented with, and are now known to work, they go in and do site preparation before they plant. In some areas where they clear, the grass comes in; they spray for the grasses, which doesn't sterilize the soil, but it knocks the grass back a year or two, giving the little seedling a chance to compete for the moisture in the so-called A-Horizon. And sometimes you might want to leave some brush around to protect from sunlight. Different shrubs take moisture from different levels of the soil. If they take it from deeper levels, you can leave them. Then after several years, you come in and spray. Or you can spray for brush beforehand and then burn, to clear the site, to prepare it, so that these little firs get the sunlight. How much sunlight they get makes a terrific difference in the rate of growth.

"Fertilization is a part of this, too. In the past year, for example, the same companies that are doing the herbicide spraying here fertilized 200,000 acres of Oregon forests. Fertilization of forests has become an economical thing; but you cannot fertilize unless you first control the weed species—otherwise you would fertilize the weed species along with the desired species. That's another reason why phenoxy herbicides are so important.

"It's unfortunate that there's misunderstanding with regard to the health issue. People who have studied it and are very well informed are not concerned about the health issue. They've considered it, and they've determined that it's not a hazard. That's the reason that the Environmental Protection Agency continues to approve its use. There is no evidence to show that it's hazardous to the human animal. If there were, they would immediately take it off. [This interview was conducted shortly before the EPA withdrew its approval for 2,4,5-T.] There is no evidence of anybody having been injured through the use of phenoxy herbicides, not just this past year but in over twenty-five years of use. It's absolutely safe. I'm concerned about my own health probably more than anybody else's—I wouldn't want to hurt myself or my family—and we have sprayed in our own drainage up in Redwood Creek where we get our water supply. We've sprayed within a quarter-mile of where we take water. We've monitored the water too, and there's no sign of any spray in it.

"People have actually drunk this stuff straight, and it didn't hurt them. I've studied this stuff. I've gone down to the University of California and gone into the laboratories and talked to the people there, and gone up to Portland and talked to authorities there, and talked with local people here at Humboldt State, and all the people who are really well informed say that there's absolutely no hazard or risk with it. You can go out here to the drug store right now, and go in the garden department and buy this stuff right off the shelf in a pint bottle. You can read the label on that: it has 2,4-D and 2,4,5-T. It's exactly the same stuff. It's been out for years. Millions of people each year work with that stuff.

"The irony is that when we go out and put it on the forest with a helicopter, we put it on in the lightest possible dose to do an effective job. There are professionals all the way through: professional applicators; professional pilots; professionals from the Department of Agriculture monitoring it; and our own licensed foresters monitoring the water, watching for the first sign of drift. You can see the material—if it drifts, you can see it. Then you'd stop if it got too windy. They have very strict requirements. So with all of that, it seems to me that it's obvious and apparent that it couldn't really be dangerous. Doesn't it to you?

——— " ———

Mechanization Takes Command: Technological Delights

The final step in intensive forest management is the harvesting of timber. But this is not really the end, for the cycle will be repeated indefinitely: collect seeds; nurture seedlings; prepare the ground; plant; spray to eliminate weeds, insects, or disease; thin the rows of young crop trees; fertilize the ground; thin again; and, finally, harvest. The process cannot be repeated every year as with food crops, but on most sites under intensive management it can be repeated at least twice in a century.

A second-growth tree crop at harvest time differs markedly from old-growth timber. There is less rot, less breakage when falling the tree, and fewer unwanted trees to get in the way of the harvesting operations. There is also a lack of the pure, fine-grained wood that takes a century or more to develop. The fine grain is a direct result of slow growth—and slow growth can no longer be tolerated by industry. The new trees are not yet old enough to have shed all their bottom limbs, and the wood is therefore knottier. And since each tree is still growing rapidly, a high percentage of it consists of sapwood, which lacks the strength and durability of heartwood. Second-growth redwood timber, for instance yields only 5 to 10% in clear grades of lumber (free of knots and sap), while 60% of the boards from old-growth redwood are clear.[9]

As wood quality changes, markets must be found to adapt to the new products. The California Redwood Association, a marketing organization, once advertised only their clear and construction grades of lumber; today, they emphasize the use of sapwood for paneled interiors and knotty lumber for outside patios, arbors, and fences. The mills, too, must adapt to new-growth trees: they scale down their equipment to deal with the smaller logs, and they process a larger percentage of pressed and glued products to make use of even the lowest grades of wood.

Since the new crop of trees is of uniform age, all the logs are approximately the same size. This regularity makes it feasible to develop new machinery specifically geared to the harvesting, handling, and processing of even-age timber. If the terrain is gentle enough, the entire harvesting operation can be done mechanically: a giant pair of shears snips off the trunk at its base; the tree is immediately rolled through a moveable delimbing and bucking machine; large claws called grapples pick up the logs and load them onto trucks. Hand-operated chain saws and cable rigging become obsolete, while timber fallers, buckers, and choker setters are replaced by heavy-equipment operators working at control panels within the cabs of their machines. There is no need to hire real people to remove the numerous branches from the bushy trees, or to tie the many small stems together for yarding, when machines can to it faster and cheaper.

This sort of mechanized harvesting show is becoming commonplace on tree farms throughout the South. Many of the rugged hills of the Northwest, however, inhibit the movement of such large-scale equipment, so this area is mechanizing more slowly. Even so, the second-growth stands in the West that are ready to harvest tend to be on relatively flat and accessible terrain, since those were the areas that were logged first. One study concluded that 60% of the forested land in western Oregon, and 85% in eastern Oregon, is capable of being harvested by machines.[10]

Intensive forest management generates its own technology. There are several specialized machines, for instance, designed exclusively for crushing slash and brush during site preparation: the Case Tree-Eater, the Young Tomahawk, and the Kershaw Klear-Way. Hydro-Ax has developed three attachments to fit on a single tractor: a feller-buncher, which cuts and piles the trees; a chain flail delimber, which removes branches and turns them into mulch; and a brush cutter for site preparation and precommercial thinning. For areas in which tractors and other heavy equipment cannot operate, the slash and brush can be crushed to the ground by a giant steel cylinder filled with concrete and suspended from a high-lead cable.

Explosives can also be used for brush clearing: a canister of propylene oxide is detonated to produce a shock wave that literally strips the leaves and branches from the plants. The frequent use of burning for site preparation has triggered the development of an impressive arsenal of fire ignition devices: shotgun tracer shells, napalm grenades, and electrically detonated firebombs. Aircraft are utilized for a variety of purposes: to ignite and put out fires; to spray herbicides, pesticides, and fertilizers;

to survey the land and cruise the timber; and even to harvest the logs.

This increased reliance on technology has created a new system of values among loggers. Turning trees into lumber is still seen as a challenge by which a logger can assert his manhood, but it is no longer the physical prowess of the lumberjack himself that leads to success on the job. The machine has come to man's aid: it is the extension of his own brute force, the final realization of his control over the forest. Advertisements for the

Tree tongs: a hydraulic grapple loader

tools of modern logging reveal the psychological equation of heavy equipment with personal strength:

International: Logger's Word for Tough

If you're moving neat little boxes down smooth and easy highways, maybe you can settle for less. But when you're into logging, you've got to move up to International.

Big loads and back roads—or no roads at all—that's where the long-nosed brute of the woods comes into its own. The Transtar Conventional levels hills as easily as a chain saw cuts through kitchen matches; hauls loads that'd shock Paul Bunyan.[11]

Paul Bunyan's Blue Ox has been reincarnated as a modern logging truck with a full payload. His ax has become giant scissors that snip the trees from their stumps. His hands are the grapple hooks that make logs seem like toothpicks. So where is Paul Bunyan himself? He flies high overhead, surveying his plantation with aerial photography. He sits in the cab of his fully mechanized harvester. He programs a computer to manage his scientifically organized tree farm. The task he faces is not how to fall and move a ten-ton tree, but how to grow ten one-ton trees in its place.

3

TREE SAVING
The Voice of Ecology

There is more to the woods, say the ecologists, than resources that are produced, processed, and consumed. The forest exists not just for human use; it has a life of its own as well. The forest is a home for creatures of the wild. It is a symbol of a vitality that goes beyond anything we could create ourselves. Yet now the forest is being tamed: the engineered trees, all planted in rows, are like animals raised in a zoo. As lifetime timber faller Bob Ziak, Jr., puts it:

> I really don't think there are going to be any forests of the future. Forests to me mean splendid old trees, with animals in them, with a variety of trees, snags, and windfalls: everything that it took to create over the hundreds of years that the live trees stood. I feel certain that the forests of the future which the big companies talk about now are going to be nothing more than what they have begun to call them: tree farms. And you won't be able to walk in them because a tree that is thirty-five years of age, he's still bushy, he's still got a lot of little pin limbs. There'll be no majesty; there'll be no cathedral-like feeling.[1]

We ourselves are products of the forest. Important human physical characteristics—binocular vision and prehensile hands—were developed as adaptations to life in the trees. Over time, however, *Homo sapiens* left the woods: less than 1% of the earth's population currently lives under a forest canopy. Still, we maintain a biological relationship with the forest environment. In the words of ecologist Esteban de la Puente:

> Healthy forests are an important part of our birthright. Forests fix almost half the total energy of the biosphere. When sunlight

falls on the forest, radiant energy is transformed into chemical energy. Photosynthesis turns carbon dioxide from the atmosphere into carbon compounds—the substance of living organisms. In this process of chemical digestion, the forest vegetation releases oxygen back into the air. We, in turn, utilize the oxygen and send back carbon dioxide as a waste product. It's a neat bit of recycling. We coexist. We feed each other with our respective wastes. We're organically linked—the forest and ourselves.[2]

Ecologists perceive a forest not as a farm or a factory, but as a complex biological community with its own economic structure. Ecology—the study of the economics of natural systems—focuses on the interconnectedness of all organisms in a given environment. Over millions of years, a process of mutual adaptation has developed by which each species within the forest community coexists with its neighbors. Today we are trying to reshape this forest ecology, simplifying its structure to serve our own needs. Our real needs, however, cannot be measured in board feet alone. According to the opponents of industrial forestry, our technological manipulation of the ecosystem threatens to deprive the forest of its own economic viability. Unless we change our ways, they claim, we look forward to a spiral of diminishing returns, and ultimately, perhaps, to total deforestation.

Gordon Robinson

Forester for the Sierra Club

"I believe in multiple-use forestry, which means modification of timber management to provide for the other uses of the forest. Multiple-use forestry consists of managing the forest within the following guidelines or parameters. First, a *sustained yield*, and for that I have my own definition: sustained yield means limiting the removal of timber from a property or administrative unit to that quantity which can be removed annually in perpetuity, where the quantity may increase and the quality may improve, but neither can ever decline. The Forest Service talks about sustained yield, but they're constantly shortening the rotations and decreasing the quantity and quality to sell more timber *now*, to satisfy local, temporary demands by the timber industry.

"The second parameter for multiple-use forestry is to *practice uneven-age management in preference to even-age management, logging only under a selection system.* That means keeping the openings in the forest resulting from logging no larger than necessary to meet the biological requirements for regeneration. There are several species that require open sunlight to reproduce and grow satisfactorily, but that information should not be used to justify cutting forty acres at a crack. You don't have to clear forty acres to let the sun hit the earth. The sun's hitting the ground right here in this yard, and this is only a tenth of an acre.

"The third parameter is to *allow the dominant and codominant trees to mature before cutting them.* Mature in this sense means allowing the trees to achieve their full height. Trees grow like people: they grow up, and then they grow out. Trees have pointed tops, but after a while the upper limbs flatten out, and that is when you call the tree mature.

"The fourth parameter is to *maintain the habitat for all of the species of plants and animals that live in the area.* The reason for this is that there are many subtle interrelationships among the species that live in the forest. We don't know very much about it, but we know enough to know that they are important.

"Take the woodpecker, for example. Woodpeckers are insect-eating birds, as you well know. They're very noisy eaters. Well, woodpeckers need old trees, generally ones with broken tops and decayed hearts, for nesting. It's necessary to have some of these old snags standing as habitat for the woodpeckers if you're going to have them around; and woodpeckers are very important as a control over the bark beetles that kill the trees. I think that the present epidemic of bark beetles in the southern pine region is clearly the result of having eliminated the habitat for the woodpeckers, which reduced the woodpecker population to next to nothing. Not only woodpeckers, but there's a series of species that use these holes. Woodpeckers build them, nest in them, and move on. Then somebody else will move in. These apartments are rented; the tenants change from time to time. But the birds that live in these nests are insect-eating birds, so it's necessary to maintain the habitat for all of the species that occur naturally in the area .

"The fifth parameter for multiple-use forestry is to *take extreme caution to protect the soil.* If we're going to preserve the productivity of the land, we have to protect the soil. When you remove all the vegetation, you expose the whole area to the leaching of nutrients and the erosion of the soil. Soil scientists talk about two basic types of erosion: sheet erosion and gully erosion. There's a tendency to assume that there's no erosion unless you see deep gullies. But sheet erosion is where a whole layer is moved off the surface. The signs are little pebbles spread over the surface: the earth has washed away and left those pebbles. Well, large quantities of earth are lost to sheet erosion following clearcutting. It varies, of course, according to the steepness of the slope and the texture of the soil, but there's some loss anywhere, and frequently there's a lot of loss.

"Another problem is compaction. Going over the land with all this heavy equipment packs it down. Soil is composed, generally speaking, of 50% mineral, 25% water, and 25% gas. If you pack it down and squeeze the gas out of it, you destroy the environment for the creatures that are living off each other in there. In one square foot of earth, there is a population of about 10,000 individuals divided into somewhere between 50 and 200 species. They're all different sizes, and the big ones eat the little ones, just like the fish in the sea. The most densely populated zone in the biosphere is the top foot of earth. Now if you pack that down tight, as you do when you drive over it with tractors or mash it down with these machines which chop up the brush, or you scrape it off with bulldozers and re-pile it, it's a disaster for billions of individuals.

"Those individuals have all kinds of different functions. There are indications that some of the fungi are able to extract mineral ions out of the large rock, the solid mineral component of the earth. We don't know how phosphorus, for instance, gets out of the rock and into an organic compound, where it becomes available to plants and animals. I think it must be the function of microorganisms. This is a vast area of research that people are just beginning to look at.

"There are some general things that we do know about these interrelationships. The roots of conifers tend to be coarse and blunt compared with those of other plants, and they don't have root hairs. So conifers on their own cannot successfully compete with other seed-bearing plants; but different species of mushroom penetrate the roots of the conifers and draw on the trees for their sustenance: synthesized sugars and starches. Then the mushrooms, in turn, supply the mineral nutrients to the tree. It's a symbiotic relationship called mycorrhizal, and the mushrooms that do this are mycorrhizae. We don't know exactly how these guys work, but we know they're there and we know they're interrelated, and it behooves us to protect them if we're going to survive on this planet.

"Now the mycorrhizae can multiply very rapidly, and I suppose that nature eventually repairs the damage we cause by our logging. But the big question is: can we permit practices that accelerate the rate of destruction? Erosion is a natural process, as foresters will tell you, and they'll use that as an argument to justify what they're doing. But it doesn't justify it. There is a natural balance between the rate of formation of soil and the rate of natural erosion and destruction of soil. We've got to maintain that balance, or continually reduce the productive capacity of the earth as we increase our own population.

"And we've got to maintain the balance of the various species. The combinations of species that naturally exist are together because they need each other, because they survived under trying circumstances. I think we can assume that each element plays a role in the common survival of the ecosystems. I don't think we can ignore the hazards of monoculture in the hope of maximizing income and profit. I think that's a delusion.

———— " ————

Strength in Diversity

One of the basic laws of ecology states that: *Other things being equal, the degree of environmental stability is in direct proportion to the number of species living together in the environment.*[3] Insofar as monoculture simplifies the ecosystem by decreasing the variety of species, it would seem to lead to increased ecological vulnerability.

Disease, of course, presents a constant danger to the life of the forest. Tree diseases are many and varied, but most are specific to certain species. When a large number of trees belonging to a given species are concentrated in one area, the spread of any disease which affects that species is facilitated by the dense population of susceptible trees. Epidemics, therefore, are more likely to occur in pure stands than in mixed stands. According to John R. Parmeter, Jr., professor of plant pathology at the University of California:

> One of the cardinal principles of plant pathology is that the greater the purity and density of a plant species, the greater the likelihood of serious plant damage. Forest pathologists long have called attention to the dangers of monoculture and to the desirability of good species mixtures. . . . The factor of density

dependence in disease epidemiology is so generally applicable
that it may well be one of the main ecological mechanisms
driving plant communities toward the stability of diversity.[4]

In other words, one of the reasons that pure stands of a single species are
rarely found in nature is that susceptibility to disease has resulted in their
natural elimination.

Insects also tend to focus on specific types of trees. Most insects use a
single species or a small group of species as their hosts. Even insects that
are general feeders tend to prefer the host species upon which they were
reared. In a pure stand of a single tree species, insects which require or
prefer that type of tree can find a readily accessible and virtually limitless
supply of food. According to a textbook on forest entomology:

> It is an accepted biologic law that, other things being favor-
> able, an organism will eventually multiply to the limit of its
> food supply. As a rule, the more numerous the individuals of a
> tree species, the more abundant are its insect enemies. When
> there is an unlimited and convenient supply of a certain species
> of tree, the stage is set for an outbreak of the insect pests of that
> tree.[5]

The southern pine bark beetle, for instance, thrives on the planted
monocultures which have replaced the mixed forests of the Deep
South.[6] And since host trees are often most vulnerable at specific stages
of development, the even-aged tree farms are even more susceptible to
unimpeded outbreaks of pests.

In a diverse environment, on the other hand, there are natural limitations on insect populations. The insects have to search farther afield for
their food; whether or not they find it, they expose themselves to natural
predators in the process. Tree species that generally occur in thoroughly
mixed stands are therefore considered more insect resistant than those
that occur in near-pure stands. Jack pine, when it occurred only in scattered blocks on poor soil, was once thought to be an insect-resistant
species. Recently, however, jack pines in heavy concentration have
replaced the virgin forest species in the Great Lakes states, and these
trees are now subjected to repeated attacks by various pests. Jack pine is
no longer considered a highly resistant species.

The agricultural analogy that industrial foresters like to draw is helpful
in understanding the problems with tree pests in monoculture. Many
insects that feed upon domesticated crops—such as the Colorado

potato beetle—existed long before the land was tilled. When the vegetation varied, the insects were few and far between. Food in any given area was scarce, so the beetles were forced to travel to new locations where they might or might not find sustenance, and where they would be subject to natural predation. When extensive potato fields were planted, however, the beetles from the surrounding areas suddenly found an abundant source of food that could be had for the taking; they multiplied wildly, beyond the control of natural predators. To combat the pests, new strategies had to be developed, and so it was that the spraying of pesticides came to be a standard agricultural practice.

As pure stands of trees come to dominate the woods, foresters are starting to face a similar situation. Losses to insects and disease on today's tree farms are not that great—but only because the pests are now controlled with the strategic application of chemicals. The danger is not that infestations and epidemics will wipe out all of our young trees in the near future, but rather that the forests will become dependent on man-made props. Natural balancing mechanisms, such as varying stand composition and encouraging populations of insect-eating birds, are bypassed. The prevention of infestations and epidemics is replaced by artificial controls. In the words of Frederick E. Smith, professor of Resources and Ecology at Harvard:

> The use of pesticides may seem necessary, even though the manager is aware that it replaces, rather than supplements, the effects of predators and parasites. From then on the system enters upon a long spiral of degradation, as the course of intensive agriculture clearly shows. Agriculture is successful today only because an enormous input of power and materials is used to keep an increasingly unstable system in line. . . . The most perilous aspect of pesticide use is the addiction that follows repeated use. One guaranteed consequence of using pesticides is an increased need to use them again. Not only do pests tend to recover from treatment faster than their enemies, but additional pests are created as other predators and parasites are inadvertently damaged.[7]

Technological Addictions

The dependence on engineered controls constitutes a feedback system: as some controls are used, further controls become necessary. The use of pesticides is only one example; there are several other ways in which our forests are becoming increasingly dependent on human support. When we fertilize the soil, for instance, we supply whatever element has been the limiting factor on growth. The addition of this element, however, enables the trees to utilize more of the other nutrients they need, and the soil is thus depleted of other essential elements at an increased rate. Increased growth is achieved for the present crop of trees, but there will be fewer nutrients available for future crops, and hence even more fertilizers will have to be used.[8]

Another example is the destruction of mycorrhizal fungi during the harvesting of timber. Mycorrhizae aid young conifers in several important ways: (1) they break down minerals from the soil into forms which can be utilized by the trees; (2) they increase the surface area of their host roots, which can therefore absorb more moisture; (3) they contribute to the production of humus; (4) they secrete organic "glues," stabilizing the soil structure and permitting the movement of air and water required by plant roots and other soil organisms; and (5) they protect trees against soil pathogens by providing physical barriers against penetration and by excreting antibiotics and other pathogen-inhibiting organisms.[9] Clearly, mycorrhizae pay a fair price for the photosynthetic energy they receive from their hosts.

Unfortunately, however, clearcutting and prescribed burning have a severe impact on mycorrhizae populations. Harvesting all the trees eliminates the hosts, depriving the mycorrhizae of their source of energy. Although new hosts will appear on the site in due time, the conditions conducive to the survival of mycorrhizae are altered by clearcutting and burning: soil temperature is raised, pH is higher, litter and duff levels (where mycorrhizae are most often found) are diminished. The combined effect of host elimination and environmental change is a decrease in mycorrhizae. A study of 36 "difficult to regenerate" sites in northwest California and southwest Oregon compared the number of mycorrhizae on seedlings grown in soils taken from (a) areas which had been clearcut and burned, (b) areas which had been clearcut but not burned, and (c) nearby undisturbed areas. The soil from clearcut and burned areas produced 40% fewer mycorrhizae than soil from the undisturbed areas, while the clearcut but unburned soils produced 20% fewer mycorrhizae. This reduction of mycorrhizae seems to be a contributing factor to regeneration failures on many sites.[10]

As foresters have become more aware of the important role of mycorrhizae, they have started to inject the soil in tree nurseries with the appropriate spores. Ironically, the routine fumigation and high fertility

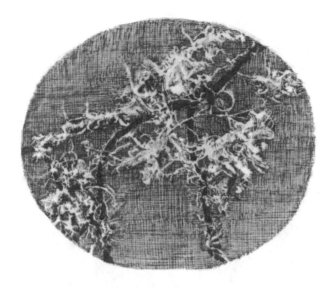

Douglas-fir root hairs with mycorrhizal fungi (magnified 5 times)

levels in nurseries have historically impeded the survival of mycorrhizae, which are now being purposely reintroduced. Since management techniques have impacted the mycorrhizae in several different ways, forest managers are now trying to make amends by replacing what they have just removed. But are their replacements adequate substitutes? Will the shots given in the nurseries work as well as the diverse and locally adapted mycorrhizae which are found out in the woods?[11]

After natural catastrophes in which the hosts have been removed, mycorrhizae recolonize an area with the help of small mammals such as squirrels, mice, and voles. These creatures of the soil feed upon the underground reproductive bodies (called truffles), passing the spores through their digestive tracts and spreading them generously from the nearby forest to the disturbed areas. But small mammals have long been regarded as enemies of reforestation, since they also feed upon the seeds and seedlings of commercial tree crops. Often, they are systematically destroyed, either directly or by eliminating their habitat. (Red-backed voles and deer mice, for instance, live in rotting logs, which historically have been removed in the wake of clearcutting.) With fewer creatures around to spread the truffles, the availability of natural reproductive

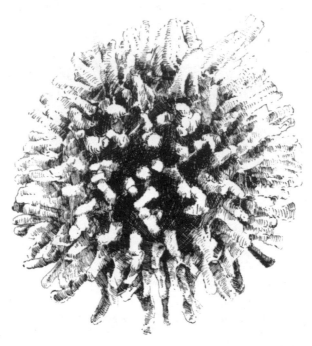

A single spore from the fruiting body of a mycorrhizal fungus (magnified 3,000 times)

mechanisms is reduced. The ecosystem is being simplified by limiting the numbers of animal "pests," while an additional step is being added to engineered reforestation in order to compensate for the function these animals once served.[12]

The simplification of the ecosystem is particularly dangerous in the field of genetics. Man, not nature, can now select which seeds will grow into trees and which will not. The seeds that man selects come from a relatively small sampling of parent trees. Variation becomes more limited when a few hundred trees in a seed orchard bear the entire burden of reproduction for many thousands of acres of forested land. Yet it was natural variation that created the giant, straight-grained conifers so highly prized by the timber industry. In nature, only the hardiest seeds are even germinated; in the nursery, almost every seed becomes a tree. In nature, only the strongest and the best-adapted seedlings grow to maturity; under human control, profitability rather than endurance serves as the key to selection—and there is no guarantee that the fast-growing trees will produce a genetic line that is best adapted to environmental stress.

When all the trees in the forest are of the same species, they are more vulnerable to pests and pathogens; when these trees also come from a limited gene pool, the danger is even greater. Since many pests and pathogens focus on specific genotypes, the likelihood of epidemics is increased by a narrowing of the genetic base. Ironically, the genetic base is shrinking at a time in history when all sorts of new stresses are being placed on the forest. With the global climate beginning to change, and with man-made pollutants beginning to drift toward the forests, the trees are being challenged as never before. Is this any time to sacrifice resilience for the sake of short-term gain?

Geneticists, of course, are aware of these dangers and are taking several steps to minimize them. In order to avoid excessive narrowing of the gene pool, they select trees for their seed orchards from a variety of locations and elevations. In order to test for disease resistance, they inoculate different trees with various diseases and select only the most resistant strains. In order to test for environmental adaptability, they keep close records of the progeny that come from each of the genetic strains they develop.

At best, however, all the tricks the geneticists have learned cannot match the ultimate test for environmental endurance: survival for thousands of years in a natural setting. This is nature's test, and the gene pool we have in our untouched forests represents the strains that have passed this test. Nature does not select trees according to how easily they can be run through the mill or how quickly they can produce wood fiber. Insofar as geneticists make their selections according to these new criteria, the genetic development of the trees turns away from the simple test of survival. Ultimately, this could mean further dependence on a network of human assistance to provide for the continued health and survival of the desired trees.

Again, the problem is made clear by drawing an analogy with traditional agricultural crops. After thousands of years of breeding for improved strains, most food crops today would be incapable of fending for themselves. In the words of Forest Service geneticist Roy Silen:

> The history of plant improvement is that we usually have been thorough in replacing the original gene pool with an "improved" strain. For example, by the time of Columbus, corn had been so altered by selection done by the Indians that the original wild corn plant was extinct, and corn's future was entirely dependent upon man: it could not persist in the wild. The history of improvement in wheat similarly has been the

refining of the gene pool into one strain after another, none of which could exist without man; each, in turn, was wiped out by a pest. Fortunately, a new improved strain was always in the wings, arising from some resistance gleaned out of the shrinking original gene pool. Today the original gene pool of wheat is reduced to a few small acreages in the "fertile crescent" of the Mediterranean area.[13]

During the 1960s and '70s, there was great excitement over improved strains of rice that were supposed to help solve some of the food problems of the underdeveloped countries in the Far East. But the much heralded "Green Revolution" has run into problems: the rice cannot be grown without the repeated use of fertilizers and pesticides, which the impoverished peasants can hardly afford. Could something similar to this happen to our improved strains of trees? A thousand years from now, will our trees require—rather than simply prefer—the application of fertilizers, pesticides, and herbicides? And, if so, will we be in a position to support their addiction? Can we afford to commit our forests to a perpetual dependency on artificial supports?

Of course we can only speculate about the effects of increased dependency on human engineering, and we can only guess whether foresters of the future will be able to handle the consequences of this dependency. We don't really know where our actions are leading us. But it is this very uncertainty that makes many ecologists uneasy. Nature, they say, knows best. It is presumptuous, they suggest, to think we can improve on millions of years of natural experimentation. We cannot bypass evolution.

Fred Behm

Hunter and Lumberjack

"I belong to this small tree farm organization, and a year ago we made a tour of the Weyerhaeuser tree farm at Cottage Grove. They took us up on the mountain there, and we could look down across it: it was one of the most beautiful sights I've ever seen, thousands of acres of reproduction. But then I looked at it closely and got to thinking about it. It's one of the most depressing things I've ever seen, too. Nothing is going to be there. They'll precommercial thin when the trees are fifteen or twenty years old, then in fifty years they'll clearcut. It'll be like a cornfield. There'll be no snags, no old logs in there. Wildlife—squirrels, birds, and salamanders—will have no place. So many creatures live in the snags and under the old logs. What's it gonna be like without them? Like I say, it's beautiful but depressing.

"My personal opinion is that nature knows what it's doing, and if you interfere too much with nature, you're going to have problems. In nature one thing takes care of the other, and you can't interrupt that without destroying something.

"You can't have it all be just the same. They're breeding these trees for improved yields, but I'm not too sure how that's going to turn out. Of course, they've done that with the grain and the corn and everything else, so I suppose it's possible. But you can overbreed, too. Just like with horses, you can breed up too high where you don't get a good animal. Could be that way with trees, too. But how do you know? You're looking three, four hundred years into the future, maybe a thousand years. I don't think we should put all our eggs in one basket. We should let nature take its course, too.

"I grew up in Wisconsin. I got my first ax when I was seven. I started working in the woods when I was nine. That's when I made my first day's wages. I've been working in the woods about sixty years. I enjoyed every minute of it, too. If I had it to do over again, I wouldn't change a thing.

"When I was in the sixth or seventh grade, the school superintendent came down and asked the kids what their ambitions were when they growed up. They wanted to be presidents, doctors, lawyers, and nurses. Well, we had a neighbor, a Norwegian bachelor who cut cord-wood. He was my idol. I said, 'I'd like to be a cord-wood cutter like the old man.' The superintendent said, 'Son, you must have bigger ambitions than being a cord-wood cutter.' I said, 'No, if I can be a cord-wood cutter like the old man, that's good enough for me.' I've often thought about it; I bet I hit my ambition closer than any of the others did.

"First time I came to Oregon, I saw that pretty water in Blue River and I said, 'Boy, that's the place for me.' So I went and bought a lot, and I'm still here. And I bought timberland. I had 1,700 acres at one time, but when they built the dam they took about 400 acres.

"I've been working for myself since '43, logging my own timber. I run it on a sustained yield—take out so much a year and reforest it. I'm kind of a jack-of-all-trades, so I do a lot of the work myself:

fall timber, hang the rigging, set chokers, run loader, run cat. It's my life. I enjoy it. I'm sixty-eight and I'm still doing it. I feel more comfortable working in the woods than pushing pencils.

"I like to hunt. Not so much on my own land, because I like to keep the game around. Of course, the deer are a problem: they browse the trees, the seedlings, and cause a lot of damage. But I don't begrudge it to them. There was deer here before I was. The timber always did survive the deer browsing on it.

"I prefer to hunt up in the high country, up in the Three Sisters Wilderness Area. Pretty country up there. I just hunt for the fun of it. Lot of times, I don't try to shoot anything. I just go around and see what I could get if I wanted to. One year I turned down thirteen bucks. Never did kill one that year. Just looking for a bigger one all the time. Couple of years I carried my bow and arrow with me instead of a rifle. And I do a lot of camera hunting. Several years that's all I did was take moving pictures. I get more of a kick out of that.

"Personally, I think we should keep some land as wilderness. It's important for wild game. I don't think it should all be managed like a tree farm. I'm very disturbed with this idea of trying to cut off all the old growth. I think they should leave a certain amount, especially in areas that are winter range for elk and deer. In a bad winter, that's how elk and deer survive, off that old growth. There's a lichen on the old growth—they just made a study on that down at the university—and it produces nitrogen, about 300 pounds per acre, I think it is. Well, that same lichen is feed for game during a real bad winter. That's nature's way of feeding them: the snow breaks the limbs off that have got that lichen on there, and that's good food for deer and elk. The heavier the snow, the more limbs break off. The brush is covered up on the clearcuts; it's completely snowed over. You get four or five feet of snow and you won't find an elk or a deer track out on a clearcut unit. They'll all be up in the old-growth timber stands.

"Trees don't grow that lichen until they're about ninety to a hundred years old. So if we go with a rapid yield like Weyerhaeuser and these companies are thinking of doing, I don't know. You can raise more board feet by the tree farm method, but it shouldn't all be that way. They should have blocks of old growth and have it perpetual, so that when one crop of old growth gets to a climax stage, you have another block coming to take its place.

"I still have patches of old growth on my land. When they reach a climax stage, they're better off cut, because all they do is deteriorate. I'll let them grow two or three hundred years to reach a climax. I hope I can do it that way. If I get broke I may have to cut them.

"You can't just look at today, you've got to think ahead. That's the reason so much of our private land doesn't have trees on it: people just can't see that they've got to look that far ahead. It's not like growing corn, where you get a crop every year. Weyerhaeuser, they figure fifty years, but that's kind of stretching it a bit. They're figuring out how many cubic feet they can get out of an acre every year. That may be right, but then they're going to have to fertilize, and spray, and all this sort of stuff.

"I don't think these big companies are looking far enough ahead. In the long range, what's going to happen to the soil? In any kind of farming, you've got to have something to build it back up again or the soil goes to pot. Most of these fertilizers they're putting on are just a feed, not really a fertilizer. They're just short-term, a shot to make things grow. They don't really build the soil up. For that you've got to have humus, a certain amount of needles, and stuff like that. You've got to look way into the future, not just fifty years.

———— 〃 ————

Overdrawn at the Bank: Soil Depletion

In the long run, the most important single resource of the forest is the soil itself. It is the soil that must provide nourishment for tomorrow's timber. Soil is the fragile skin of the earth, the interface between organic and inorganic elements that constitutes the life-support system of the forest. It takes hundreds of years to create an inch of topsoil by the weathering of parent rock material and the decomposition of humus. In a mature forest, the cycling of soil nutrients is slow and balanced: what is used up by the trees is replaced by natural fertilizers such as rainfall, rock weathering, and the litter that the trees themselves create. Under natural circumstances, the soil itself endures for periods of time which are difficult for us to comprehend; the residence time for a particle of soil in the Oregon Cascades, for instance, is about 10,000 years.[14]

Soil builds slowly, but it can go fast. Nutrients can be removed from the forest in a variety of ways. They can be volatilized, released into the atmosphere. They can be leached from the topsoil back into the rock mantle. They can be dissolved or suspended in water and carried away by streams. Or the organic biomass of the forest can simply be transported off to other locations.

Intensive timber management practices create nutrient losses in each of these ways. The controlled burning used in site preparation turns nutrients into gaseous forms that are released into the air. Clearcutting exposes the soil and increases the quantity of water that flows through it, thereby magnifying the extent of leaching. Road building and other earth-moving activities result in erosion, causing the soil itself to be transported into the rivers and out of the forest. And logging, of course, reduces the forest biomass, thereby lessening the amount of organic material that is available for future soil production.

The most natural tool used in timber management is fire. Long before the evolution of human beings, fires caused by lightning were thinning out forests and controlling insects. Yet even these natural fires left an impact on soil nutrients. Nitrogen, the basis of plant protein, is the nutrient most directly linked to tree growth—yet nitrogen literally disappears into thin air when a wildfire consumes the organic material on the forest floor.

Ironically, although the nitrogen bank of the soil is depleted by fire, the amount of nitrogen immediately available to young trees is temporarily increased. Since controlled burning also makes the ground more accessible for reseeding and planting, forest managers have utilized fire as a tool to facilitate reproduction. Unfortunately, however, the short-term increase in available nitrogen is offset by the long-term depletion of the nutrient bank. The precise extent of the damage from a controlled burn depends on the heat of the fire: sometimes only the litter is consumed, but at other times the organic component of the ground itself is destroyed.[15] Volatilized nitrogen ranges from 100 to almost 1,000 pounds per acre.[16] In British Columbia, a low-intensity burn released 431 pounds of nitrogen per acre into the atmosphere, while an intermediate-intensity burn released 876 pounds per acre.[17] When piles of logging slash are burned, the heat is intense enough to volatilize over 90% of the nitrogen.[18]

Fire also destroys the microorganisms that contribute to the breakdown of soil nutrients. Microbial activity and mycorrhizae populations are highest in humus and decaying wood—yet these are consumed by intense fires.[19] If hot enough, a fire can even reduce the wettability of the ground; the soil literally repels water for 5–10 years.[20] And when rain runs off the surface of the earth instead of penetrating it, more soil is likely to be lost to erosion.

Nature, of course, has ways of dealing with these problems. Whether a fire is created by lightning, careless campers, or forest managers is of little significance. As soon as an area is burned, a natural scheme to regenerate the health of the soil is instantly set into action. Fire converts litter and humus into ash, which has a higher pH factor; this decrease in acidity encourages the growth of bacteria that are able to take nitrogen from the air and make it available to plants. Although the nitrogen bank has been depleted, the production of nitrogen is stimulated as the forest begins to build itself back.

The heat of the fire opens many types of seeds that would otherwise have remained dormant, and brush that has adapted itself to a fire ecology quickly comes to dominate the landscape. Many of these pioneer species, such as the several varieties of *Ceanothus*, are equipped to fix atmospheric nitrogen into the soil.[21] The roots stabilize the exposed ground and help prevent erosion; root penetration also helps break up compacted soils. When the pioneer plants die off years later, they leave the soil enriched and aerated by their elaborate network of root channels. Organic material, created by the photosynthesis of these fast-growing plants, gradually decays to replenish the nutrient bank of the earth.

When the pioneer stage of forest succession is bypassed by the application of brush-killing herbicides, the soil is not given the time to go through its normal recovery cycle. By eliminating the "weeds," the soil is deprived of natural fertilizers—which means that more man-made fertilizers will have to be applied in the future. Alder, the ubiquitous weed tree of the Pacific Northwest, fixes up to 300 pounds of atmospheric nitrogen per acre per year, with the exact amount depending on the age and density of the stand and the amount of nitrogen already in the soil.[22] The leaves that alder trees shed each fall are an additional source of nutrients: the litter in a mature alder forest contains about 100 pounds of nitrogen per acre per year, whereas the litter in a coniferous forest contains less than one-third that amount.[23]

What do these numbers mean? Commercial applications of nitrogen range from about 100 to 500 pounds per acre, but these are only applied once or twice in a generation. Alder trees, working year after year, contribute much more than that. One study concluded that alders, through the combined action of nitrogen fixation and litter, enriched the soil by a yearly average of 124 pounds of nitrogen per acre over a forty-year period. By the end of that time, the soil under the alder trees had 4,960 pounds more nitrogen per acre than the control area under conifers.[24] Alders might be commercially inferior to conifers, but with respect to the health of the soil they are much more valuable. By suppressing these natural fertilizers, herbicides are depriving the earth of nitrogen, effectively depleting the forest soils.

Even pioneers which do not fix atmospheric nitrogen contribute to the future productivity of the soil. In California and the Pacific Northwest, hardwood "weeds" such as tan oak, madrone, and manzanita share many of the same species of mycorrhizae as commercial crops of conifers. When logging or burning eliminates the conifer hosts, the mycorrhizae would be severely impacted were it not for hardwood pioneers, which serve as hosts until the conifers can regenerate. Particularly on hot, drought-prone sites which are difficult to regenerate, there is only a small "window of opportunity" during which conditions are favorable for conifer seedlings to survive. The availability of mycorrhizae on these sites is crucial, and the presence of hardwood pioneers can form a "biological time bridge" across the "mycorrhizae gap" between generations of conifers.[25]

To test the importance of hardwoods for conifer regeneration, two scientists collected soil from sites in southwestern Oregon which had been clearcut and burned five years previously and had not regenerated well. In a greenhouse, they raised seedlings in soil which had been collected at various distances from scattered hardwood pioneers. After five months, the seedlings grown in soil collected near the hardwoods were 60% taller and over twice as heavy as those grown in soil collected over four meters from the nearest hardwoods—and they also had almost twice as many roots with mycorrhizae. Since the experiment was conducted in a greenhouse, the differences in seedling growth were not dependent on environmental circumstances. The study concluded that pioneering hardwoods induce biological activity which is favorable to the establishment and growth of conifer seedlings.[26]

Biological activity plays an important role in nature's recovery scheme. After most wildfires, some large, woody debris is left unconsumed. These rotting logs function as islands of biological activity where mycorrhizae can survive until the area becomes revegetated. When moisture is limited, as it is for many sites which are difficult to regenerate, more biological activity occurs within rotting logs than in humus, which typically provides the most active sites for mycorrhizae formation.[27] Rotting logs also contribute to nitrogen cycling, and they maintain up to 25 times more moisture than the surrounding soil.[28] When large, woody debris is piled and burned during site preparation, an important component of the recovery system is removed.

The combined effect of burning and the application of herbicides is to eliminate the "healing" stage of natural succession. Litter, rotting logs, and pioneer plants are seen as nuisances rather than important components of an ecosystem which is constantly changing with time. When these natural mechanisms for revival are bypassed, forest managers are faced with a difficult choice: they either commit the new forests to a dependence on chemical fertilizers, or they permit the gradual depletion of the soil. According to forest ecologists, neither alternative is acceptable.

Clearcutting, like controlled burning and the application of herbicides, tends to draw against the nutrient bank of the forest. In a clearcut, all the standing vegetation is cut down. The logs that are removed from the woods can no longer be recycled back into the soil. The debris left behind is generally burned, releasing many nutrients into the air rather than returning them to the earth. The forest biomass is eliminated in one stroke, and the ground is deprived of organic material to convert into topsoil.[29]

Clearcutting also affects the microclimate of the soil. In a mature forest, water is taken from the ground by the roots of the trees and transpired into the atmosphere through the leaves or needles. When the forest cover is removed, less water is lost through transpiration, which means there is more water left in the ground. But the wetter soil is a mixed blessing. With more water filtering through it, soil losses due to nutrient leaching are increased. Normally these losses would be offset by the nutritive value of the rainfall, but the collection of nutrients from the rain is largely dependent upon root surfaces, and the roots are no longer alive.

Although the ground as a whole is wetter, the surface layer of the soil may actually be drier after a clearcut. Unshaded by trees and unprotected by brush and litter, the exposed ground is subjected to the direct rays of the sun and the blowing of the wind, and surface evaporation is therefore increased. There are no leaves or needles to collect fog from the air during summer drought. The ground becomes hotter during the day and colder at night, wetter during the rains but drier when the sun shines upon it. These greater extremes, in turn, alter the habitat for the microorganisms that normally live in the soil. With fewer microorganisms to help break up the ground, the process of topsoil formation is impeded.

Slipsliding Away: Soil Erosion

With neither trees, brush, nor litter to shelter the ground, surface runoff is greatly increased. The earth is often sealed by a clay-like film which tends to repel water. When rain hits the ground, it runs downhill rather than permeating the soil. In areas of well-established, undisturbed vegetation, surface runoff is rarely more than 3% of the total precipitation. On denuded land, however, as much as 60% of the rainfall can be lost to runoff.[30] In an experimental watershed in Oregon, surface runoff was increased sixteen inches per year by clearcutting the hillsides.[31] And with more water running over the surface of the earth, more soil is likely to be carried away by the rains. In the words of a rancher from logged-over Briceland, California, "Every time it rains around here, a whole lot of real estate changes hands."[32]

Water acts to move dirt in two distinct ways: it detaches and transports small soil particles; and it lubricates the ground to initiate mass earth movements such as landslides. During a heavy storm on bare ground, raindrops can splash more than 100 tons of soil per acre into the air.[33] Some of these detached particles are carried away by the flowing water. As the water rolls off the hillside, it eats away at the earth's surface (sheet erosion) and digs out drainage channels (rill and gully erosion). Erosion due to soil detachment rarely accounts for more than 20% of the sediment that is carried away by streams, but the soil lost is especially valuable. It comes from the uppermost crust of the earth, the rich layer of humus and topsoil that is essential for the growth of any type of vegetation.

The bulk of the sediment in the streams is produced by mass movements. Steep slopes become saturated with water in the wake of logging. The earth has lost the root systems that once held it together. Hillsides at the limit of their "angle of repose" are no longer able to hold themselves up. They literally crumble, creating landslides. And landslides tend to perpetuate themselves: the slopes temporarily become even steeper, and all vegetation that might have served as a cohesive force is carried away or buried by the moving earth.

When the landslides reach the streams, the earth material either adds to the flow of water or gets deposited on the streambed, thereby raising the stream channel. In either case, the water level in the stream becomes higher, and the higher water tends to undercut steep slopes downstream from the original landslide. This, of course, creates new landslides, which deposit even more sediment in the streams. These landslides, in turn, raise the water level still higher, which may create additional problems farther downstream.

In this feedback system, the slope instabilities created on hillsides near the headwaters can have far-reaching consequences for downstream neighbors. In Redwood Creek, California, the floor of the main channel rose fourteen feet in some places due to logging practices upstream.[34] As the channel became higher and wider, it undercut the banks of the newly created Redwood National Park. Trees that had survived flooding for centuries now tumbled into the stream.

The greatest contributing factor to landslides and stream sedimentation is the construction of logging roads. In three small watersheds in western Oregon, sedimentation in the area that had been clearcut without roads was 3 times that of the control area; but sedimentation in the area that had been logged with roads was 100 times more than the control.[35]

The damage from road building can occur even if the timber is never harvested. When a cut is made in the side of a hill, the normal slopes and weight distributions of the ground are altered—and there is no guarantee that the new slope of the hillside will be stable. The earth that is piled on the outside of the road lacks any vegetative support; the removal of dirt from the inside can undercut the bank and cause it to collapse. Roads disrupt normal drainage patterns, and they can interrupt subsurface water along the inside banks. Water therefore flows where it

A new road cut: will it hold?

never did before—and it often travels over ground that has just been rendered unstable. The result, of course, is increased vulnerability to landslides. A study of a logged area in the Oregon Cascades revealed that 72% of the mass earth movements during a heavy winter occurred in connection with roads, although the road rights-of-way accounted for less than 2% of the land.[36] A study of 137,500 acres in the Klamath Mountains concluded that the erosion rate from roads and landings was 100 times that of undisturbed areas.[37] Another study covering the entire Pacific Northwest concluded that a landslide was 300 times more likely to occur along a road right-of-way than in an undisturbed forest.[38]

Road building affects not only the distribution but the structure of forest soils. When the earth is packed down by tractors weighing 10,000 to 40,000 pounds, the soil becomes solid rather than porous: the small air spaces are pressed out, microbial activity is diminished, and tree roots find it more difficult to penetrate the hardened earth. Water penetration also becomes more difficult, and this loss in permeability leads once again to an increased flow of water over the earth's surface. A study in southwest Washington showed a 93% loss in permeability along skid trails.[39] In a controlled experiment, four trips over dry soil with a fully equipped logging tractor resulted in an 80% loss in permeability. When the ground was wet, the same damage was done with just a single pass of the tractor.[40] Because compaction creates impermeable soils, the use of heavy equipment in the forest tends to increase water runoff and soil erosion.

The combined effect of clearcutting, burning, road-building, and the use of heavy equipment is to increase the amount of soil that is carried away by the heavy winter rains of the Pacific Northwest. After 6% of an Oregon watershed was roaded and 25% was clearcut and burned, the suspended sediment in the streams over an 18-year period was 24 times that of a control watershed that had not been harvested.[41] The earth that should have been growing trees was in the water, destroying rather than nourishing the living creatures of the forest.

Impact Report: A Fish Story

Salmon and steelhead trout are anadromous fish that use coastal water-ways for reproductive activities, but the habitat they require for their spawning has been seriously affected by timber management practices. As the fish return from the ocean to lay their eggs, they sometimes con-front impenetrable logjams which terminate their journey before they can reach their spawning grounds. Even if their trip is successful, the

Cat in the stream: but what about the fish?

clear gravel streambeds in which they themselves were hatched may be covered with sediment from increased erosion. And, even if they do find clear gravel beds in which to spawn, their eggs may be suffocated by sediment washed into the stream during the winter.

Young fish hatched despite these problems may encounter a drastically altered environment if there are logging operations nearby. Oxidation of the limbs, twigs, needles, and bark that are left in the streams can use up most of the dissolved oxygen in the water. Juvenile salmon, placed in shallow streams saturated with logging debris from a clearcut, suffocated within forty minutes.[42] The juveniles also find fewer deep, clear pools to use for cover. Removal of the streamside vegetation can raise the water temperature more than twenty degrees,[43] and the higher temperatures tend to decrease the available oxygen still more, while simultaneously increasing the salmon's oxygen requirements. Warm water increases the salmon's sensitivity to toxic substances; hot water actually kills the fish.

It is little wonder that fish populations have been rapidly declining over the past 50 years, since logging and road building have reached into the erosion-prone headwaters where salmon and steelhead trout like to spawn. The anadromous fish runs in northern California streams were only one-third as extensive in 1970 as they were in 1940.[44] Along the South Fork of the Eel River, salmon runs averaged over 25,000 adults per year in the 14 years prior to 1952; in the 9 years that followed, the runs averaged only 8,000 per year.[45] By the mid-1980s, the populations were so threatened that the government had to limit the fishing season severely; in the early 1990s, the season was eliminated entirely in some areas.

Timber interests respond in several different ways to the charge that logging destroys fisheries. Some plead innocence, arguing that the declining populations are due to commercial fishermen on the open seas. Others admit some wrongdoing in the past, but claim that things are different now. They point to new state laws that restrict logging along the streambeds, and they promise to do their best in the future to keep the streams clear of sediment and debris. Still others, like Dave Burwell of Springfield, Oregon, openly admit that they are willing to sacrifice the fish in order to harvest the trees.

> I wiped out the creek for two miles, but the fish only lived and spawned in the lower quarter mile. So the fish fry when we clearcut. They'll come back in five years. They only live five years in the first place. So we don't destroy. We only interrupt the fish life. We say okay, fish, you can't live here for five years. So we only destroy one crop of fish—they're expendable. To get out that crop of trees, it's justifiable to eliminate one crop of fish. That makes sense, doesn't it?[46]

There is no guarantee a degraded stream will restore itself, no assurance that the fish will return once their spawning habitat has been destroyed. And it is not just the fish that are sacrificed. Several Indian tribes in the Pacific Northwest depend heavily on the fish runs for their sustenance; declining salmon and steelhead populations now threaten these traditional lifestyles. Literally millions of sport fishermen now catch fewer fish. And, of course, fishery degradation has put many commercial fishermen out of work.

Nat Bingham

Past President and Habitat Director,
Pacific Coast Federation of Fishermen's Associations

"I think that an intimate connection exists between trees and fish. It's a symbiotic relationship. The salmon need cold, clear water in the streams they spawn in. The trees hold the topsoil up on the hills and keep it from coming down in the streams and choking them up. The trees trap the water in the hillsides above the streams; all through the dry season that hill will still be releasing a little bit of water through springs and feeding the streams cold, clear water. And big, tall conifers shade the streams and keep the water temperature down. If the water temperature goes too high—like above 75°—there is fatality. Practically speaking, 70° is the upper limit, and the ideal is below 60°. The 50° range is what they really want.

"The trees are transpiring water vapor into the atmosphere, keeping the overall climate cooler and moister, and, in turn, creating a cooler environment for those fish. Salmon are descended from a species of fish that originally evolved as a response to the challenge of glaciation. All of a sudden you had vast river systems draining the glaciers as they melted, and the fish that formerly lived in the rivers grew to a tremendous size, because there were these huge rivers and giant waterfalls, ice-cold glacial waters all over the continent. As the glaciers retreated, these fish were forced to adapt by going to sea. The salmon is an Arctic fish in origin. The water needs to stay as close to that climate as possible for them to survive.

"The salmon are performing a tremendous service when they come back into these dry hills. The returning fish bring all kinds of essential trace elements back from the sea to the land. They die on the streambanks, and the predators eat them—the raccoons, and so on—and shit them out all over the woods. These trace elements get into the soil where the trees can use them. So there's an interchange going on.

"The worst damage to the fishery is done by the incredible amount of sediment that comes from a messy logging operation. The sediment chokes the eggs by being so fine-grained that water cannot percolate through the gravel. In clean gravel the water brings oxygen to the eggs, so when the fry hatch, they'll be able to survive. If the level of dissolved sediment in the stream goes too high, it can be disorienting to small fish. Adult fish can handle sediment-loaded water and navigate through it pretty well, but little fish that haven't been to the ocean sometimes get badly disoriented, displaced, and separated from their food supply during a heavy runoff, when the water is coffee-colored. They just get washed down and starved to death.

"Although gravel is better than fine sediment, there is also such a thing as too much gravel. It fills in the pools and tends to broaden the stream bed, making it more shallow. The stream begins to meander back and forth, undercut its banks, and destroy the stream bank vegetation. We call this aggradation, which is the opposite of degradation. Degradation is where you lose too much gravel so you don't have enough for spawning beds; aggradation is where you have too much gravel from landslides and road failures, destroying the character of streams. Narrow channels with deep pools become broad, freeway-like expanses.

"The combined effect of aggradation and degradation has been devastating. The Coho are particularly vulnerable. Coastal Coho creeks that might have had runs of three or four hundred to a thousand Coho are now down to two or three pairs of fish, or no fish at all. There are probably five thousand Coho spawning in the entire state of California, where once we had a million. The Chinook have also been severely impacted. The spring-run Chinook in the Salmon River (a major tributary to the Klamath) is down to a couple of hundred fish, where once it numbered about eight or nine thousand. Obviously, a river that's called the 'Salmon River' must have had salmon in it. Almost that entire basin is in Forest Service ownership, so you can see the effects of their management. It wasn't just the timber, although that was a large part of it. They also constructed extensive road networks, which they did not maintain.

"There are 60,000 jobs which have been impacted because fish habitat was destroyed. They've put the fishing industry out of business already, and everybody is crying about what's happened to the loggers and the loss of dignity, the guys who are ashamed to go in and get their food stamps. I have a hard time with all that, because this has been happening to us for almost a decade. We have not had a full fishing season on the north coast of California since 1985. Eureka and Crescent City have been virtually terminated as salmon fishing ports.

"Back in the 1970s, timber companies like Weyerhaeuser tried to create hatchery-reared fish that wouldn't even need any wild rivers. They called it ocean ranching. Their idea was not only to sell fish for profit, but to provide a supply of salmon that was not dependent on natural spawning habitat. This would have eliminated the economic reason for protecting a watershed. With ocean ranching, you'd no longer need a fishery in your streams; all you'd need is the hatcheries and the ocean. Sedimentation would cease to be an issue, because they could say: 'We're taking care of the consumer. We supply them with lumber, and now we can supply them with fish. So we don't need stringent forest practice laws anymore.'

Coho salmon

"By the end of the '80s, it became apparent that ocean ranching had failed. Even though they had vigorously tried to restrict the ocean fishery to keep the fishermen from catching what they thought of as 'their fish,' they still failed to get adequate returns to the production facilities. There was also a growing understanding that the genetic reservoir for the species lies in the natural spawning fish, which were still threatened by habitat degradation. Hatcheries take advantage of the natural fecundity of the salmon—three to four thousand eggs per female. In nature, only a few of these survive, while in the hatcheries they can get survival rates of up to 80 or 90%. This means you can get many progeny from just a few parents, which tends, in genetic terms, to reduce the 'effective population.' This can cause the gradual loss of genetic information in the DNA, which leads to the beginnings of genetic drift—mutations which are not supportive of survival. The only way to insure the continuation of salmon populations is to maintain both the genetic base and the habitat conditions that occur in the wild.

"I have a strong personal commitment to the restoration of natural salmon runs. Biologically and economically, I'm much more comfortable depending on a resource based on the natural world, rather than one based on a whole artificial construct of hatcheries and bureaucracy and everything like that. We're trying to preserve the species, so we can begin to reverse the effects of the heavy environmental degradation that's going on.

———— *"* ————

Salmon and steelhead trout are not the only species which are adversely affected by timber harvesting. Northern goshawks, marbled murrelets, red tree voles, silver-haired bats, northern flying squirrels, Olympic salamanders, tailed frogs, fishers, and literally hundreds of other bird, mammal, and invertebrate species are also threatened by the loss of habitat—not to mention the famous spotted owls. The special attention given to owls in recent years does not imply that they are somehow more important than other threatened or endangered species, but only that they are easier to count. Considered an "indicator species," spotted owls can best live and reproduce in conditions that are conducive to the survival of the various other inhabitants of old-growth ecosystems.

Opponents of industrial forestry have found strong scientific support in the concept of "biodiversity." In the words of Charles Wilkinson:

> The priority now being given to biological diversity is a response to the realization that human beings have extinguished hundreds of thousands of animal and plant species during the past century alone and that the process of extinction is continuing at the alarming rate of thousands of species per year. Maintenance of biological diversity allows for the development of new food sources; preserves gene pools for genetic engineering; preserves a broad inventory of animals, fungi, and microorganisms for biological pest control; keeps available a source of new medicines; and fulfills more abstract ethical obligations of stewardship.[47]

For many, the scientific concept of "biodiversity" is also a matter of religious belief: if these are God's creatures, how can man claim the right to destroy them?

The need to maintain the full spectrum of living organisms has been used in recent years as an argument in favor of preserving old-growth forests in their pristine state. In some minds, however, the mandate to maintain biodiversity does not automatically preclude all timber harvesting. Increasingly, logging operations are modified in order to preserve fish and wildlife habitat. Snags are left in place to serve as homes for insect-eating birds; rotting logs are left on the ground to serve as islands of biological activity; trees are left along the streambanks to provide shade for the fish. These days, nobody seriously argues that commercial conifers are the only elements in a forest ecosystem. Decisions can no longer be made without paying heed to the habitat requirements of other living creatures. The arguments are more a matter of degree: Can the fish survive if 30% of the canopy is left over the streams? Or do they need 50%? 70%? 100%?

Within the past decade, there has been a major shift of emphasis within the field of forestry. In the 1970s and early '80s, the idea of "intensive management" was widely (although not universally) accepted. The major goal of foresters, according to this notion, was to improve upon the workings of nature in order to produce more timber; their task was to hasten the stages of forest succession and to aide the crop trees by eliminating all competitors. Although water quality and habitat requirements were not ignored, they were seen more as constraints to the primary goal of timber production than as integral components of a functioning ecosystem. In recent years, however, more people have begun to question the basic tenets of intensive management. How can one element of an ecosystem be dominant, while the others are only secondary? Doesn't that contradict the very notion of "ecosystem"—that all elements are linked, that the whole is created by the interrelationships among the parts?

Today, foresters increasingly favor the notion of "ecosystem management." Timber might still be the major commercial resource, but the growing of trees is only one of many activities that occur within a forest environment. Even if the primary economic goal is the maximization of timber production, there is a growing realization that this goal cannot be achieved in the long run without the maintenance of a healthy ecosystem. The whole is more than the sum of its parts. From this perspective, attempts by commercial interests to simplify the ecosystem appear to be counterproductive, since diversity rather than simplicity leads to ecosystem survival.

In 1989, more than 50 of the most respected scientists in the field of forestry collaborated on a state-of-the-art synthesis of the latest research: *Maintaining the Long-Term Productivity of Pacific Northwest*

Forest Ecosystems. Topics included nutrient cycling, biological activity in the soil, soil structure, erosional processes, and ecological diversity. In a sense, there was nothing really new or revolutionary about this work; it simply summarized the findings of the past few decades. But the scope of the project demonstrated the direction that contemporary forestry is taking: in order for "productivity" to be "long-term," a forest must be seen as an "ecosystem" rather than a conglomeration of trees. And the workings of ecosystems are incredibly complex, even mysterious; in many ways which we still don't understand, they resist our attempts at simplification. As one of the essays concluded:

> We in forestry have tended to follow a technical heritage that emphasizes symmetry, order, cleanliness, and efficiency. Evidence is rapidly mounting that we must now foster something of a new tradition which accepts a little disorder or chaos as part of the natural order of things. . . .
>
> Achieving long-term site productivity is a much broader matter than simply maintaining fertile soils. It requires that resilience be maintained throughout forest ecosystems so they can absorb stresses. Ecological complexity is the key to such resilience. . . . Forest landscapes are under constant assault by management practices and other human impacts, local to global, planned to unforeseen. To combat uncertainty, we must retain as much ecological margin as possible.[48]

PART II

HANDS IN THE FOREST

Working It Out on the Ground

4

HOLISTIC FORESTRY

Nature, if left alone, will certainly grow trees. The impressive coniferous forests that European Americans discovered in the Pacific Northwest less than two centuries ago bear witness to nature's bounty. But the ways of nature are slow; to some even tedious. Since we want more timber from fewer forests, we grow anxious to get on with the business of growing trees. We substitute genetic engineering for evolutionary processes; we bypass natural stages of forest succession; we even cut the timber before it has reached its optimum productivity. By intensively managing the forests, we hope to straighten out the kinks in the meandering ribbon of time. The problem, however, is that we don't always know what we are doing. In the name of making improvements, we often create new obstacles.

So what should we do? Should we sit back and wait for nature alone to heal the scars that we ourselves have created? Or should we grab destiny by the wings and hitch a ride through time by altering the genetic selection of trees, eliminating unwanted brush with herbicides, and cutting the adolescent timber as soon as we can find a market for it?

Perhaps there is a middle road, some way of tinkering with the environment that complements, rather than contradicts, ecological processes. Some people call it *new forestry*, others call it *holistic forestry*—in either case, increasing numbers of forest practitioners are beginning to explore ways of working *with* natural laws rather than *against* them. Reacting to the excesses of industrial manipulation, they are attempting to manage land according to the basic ecological principles of balance and diversity. Instead of simplifying the ecosystem by eliminating unwanted elements, they prefer to manage all elements to the best advantage of the entire forest. Like industrial foresters, they are interested in the production of timber—but they also insist upon treating the forest as a complex, interdependent system with a life of its own.

These new foresters try to maintain a variety of species and ages in the ecosystem at all times. They are reluctant to remove the entire forest canopy at once, since this will alter the microclimate and drastically change the ecology of a site. Recognizing the fact that any form of logging constitutes a serious disturbance, they seek advice from the disturbances which occur naturally—most notably, fire. Historically, fire has shaped the structure of forests, leading to the multilayered assortment of trees we find in stands that are untouched by man. By paying close attention to how fires burn, these forest practitioners try to develop harvesting strategies that simulate nature's own management schemes.

In order to maintain healthy ecosystems, the advocates of new forestry pay particular attention to fish and wildlife habitat. They like to keep a scattering of snags and rotten logs, and they insist on preserving a protective canopy along the streambanks. Even on managed lands which have already been harvested, they hope to re-create connecting corridors of mature stands between ancient forest reserves, enabling a free flow of wildlife that is dependent on old-growth habitat. And yet, despite their preference for diversified ecosystems that include large trees, they still utilize the forest for human needs. They turn trees into lumber, providing jobs in the process.

Holistic forestry practitioners treat each site according to its own specific requirements. The ecosystem, although sustained in its basic form, is modified here and there to restore balance or to enable the extraction of products. In some spots, for instance, a brush cover might be maintained to stabilize the soil or replenish valuable nutrients, but just a few feet away, where the soil is in better shape, the brush might be cut back to allow room for conifer seedlings. The premise is that each place in the forest is absolutely unique, even though there are basic ecological rules that must be adhered to if the forest as a whole is to prosper.

The concerns of holistic forestry extend from the microcosm to the macrocosm. On the one hand, each individual site has its own particular needs; on the other hand, entire forests are dependent on the fate of areas that are geographically distant. We know, for example, that the water table temporarily rises and runoff increases in an area that has been clearcut. When there are no trees left standing, there can be no evaporation or transpiration from the leaves or needles, and the ground itself must therefore receive the entire rainfall. The hydrological effects of clearcutting, however, are not limited to the specific site. Since evaporation and transpiration are diminished, very little moisture is released back into the atmosphere. When a storm passes through, there is no "recycling" of water directly back into the rain clouds. Consequently, the storm will play itself out sooner when there is little or no vegetation covering the ground. A clearcut area receives more water, but that means that there is less water available to other sites in the leeward path of the storm. In this manner, extensive clearcutting in the coastal ranges of the Pacific Northwest can actually result in a lighter rainfall for the Cascades or the Sierra Nevada. Ultimately, deforestation around the world could alter the climate of the entire planet.

It's difficult to translate this sort of macrocosmic realization into practical terms. It is unlikely that a manager of a small tract in the Coast Range, for instance, will alter his plans because of the very slight effect his actions might have on some trees in the Cascades or Sierras. But holistic foresters are concerned with whole systems, and the ultimate system is the earth itself.

Gerald Myers

Holistic Forester

"My dad, granddad, and great-granddad were Douglas-fir gyppo loggers. They homesteaded Oregon about 1913. When great-grandpa and great-grandma came, they had a huge family, all boys; so they homesteaded the whole valley up there. They logged it in the teens. That was all horse logging. They didn't do badly with it. In 1939, before World War II, the land was really in good shape, because they left the trees on the ridges for seed trees, and they didn't muck around the streams. A horse just doesn't muck things up the way a cat does. After the war, the two boys that were left got into contract logging, working out all around the Northwest.

"Growing up, I went to eight different grade schools. We took off after the war with a gyppo logging outfit, up the Columbia River gorge and up in central Washington. When I was in high school, I started working with them in the summer. There's a funny cycle about working in the woods, and I'm still affected by it. You bust your ass in the spring, as soon as you can get in, as soon as the roads are hard enough to drive on. You work six days a week. You work dawn to dark, literally. I remember getting up at 3:30 in the morning in the middle of summer, when it was still dark, ready to crank the chain saw up at first light.

"We went broke. In '48 a lot of people went broke up in Washington. It was a real bad winter, started real early. Unemployment ran out about December. It wasn't welfare, like it is now. It would run out around Christmastime, and things would get pretty lean along about January and February. Which is really screwy, when you think about it. Those guys should have been planting in the winter. They might not have made as much money, but they would've kept in shape. Every spring all the loggers would go through this big number for the first few weeks. Aches and pains all over. Bad news.

"So I grew up in that—growing up in the woods and seeing them disappear. None of the loggers wanted their kids to work in the woods. I mean they all liked it, and it was all good, honest work and everything. But none of them wanted their kids to do it. It was ass-busting work with a cycle of boom and bust.

"My parents didn't want me to work in the woods either, so they trotted me off to college. I studied business at Oregon State; I was going to be a businessman in the city. That's what they wanted. And I did it: grad school at Cal, Berkeley; the army; then a systems analyst. Thirteen years in the city. Then I came home, back to the woods, in 1970.

"When I came out here, I came with a different attitude from what a lot of people have when they come from the city. First of all, I could see this land. I knew what it really looked like before the logging. And that's a peculiar kind of curse. I couldn't totally enjoy it in its present condition, but I saw its potential, what it could look like. A lot of potentially productive forest land wasn't producing anything except sediment in the creek.

"People come out from the city and think it's all very beautiful, but they don't really know how to *look* at a hill. They look at the trees and say, 'Gee, that's pretty,' but they may be looking at an area that's just been ravaged. Too many hardwoods were released when they took the old-growth fir. The cats screwed up the microhydrology. But there's still a lot of potential. The site may have been set back one productivity class, say from a III to a IV, but it can be brought back.

"The other difference between me and the people who came from the city is that I don't hate the loggers. It's real easy for the kids to come out here and say, 'Those stupid-ass loggers, those dumb rednecks.' That's bullshit. Those people were just trying to make a living. For a lot of them, it was the only way they could live out in the country. It was the only work available, and it's good, honest, hard work. And what were they doing? They were filling a demand. It wasn't the loggers' fault they were ripping off the woods. The cities wanted the wood. L.A. was built with the wood out of this watershed here, and from Salmon Creek, and all around here.

"It's almost as if our civilization can't afford forests. We have to cut them down, get rid of them. It's not a new story. Forests are the fastest-disappearing ecosystem on this planet. There's a lot of wealth there, and it's real easy to exploit it. But once you exploit it, if you don't take some of the wealth and plow it back in, as with farming or any continuous operation, it's going to decline in productivity. It's going to go back in ecosystem succession from coniferous forest to hardwoods to brush to brush-burn rocks to sandstone to desert.

"It's happening in Pakistan. It's already happened in Lebanon; the cedars of Lebanon of biblical times are gone. Northern Africa used to be forest. They took the forest off in the early Egyptian dynasty. It was probably all gone by about 1000 B.C. Then they grazed the hell out of it with goats, and now it's desert. Yugoslavia was all forested, and it was deforested in a very short time about 280 B.C. to build the Roman fleet. All the soil that built up over eons washed down into the valleys, and Yugoslavia is still deforested. Rome and the hills around Italy: the same thing. All around the Mediterranean: the same thing. The cycle is clear.

"People who are not able to learn from history are condemned to repeat it. We're seeing that right here in California. You start with a healthy forest on this sedimentary sandstone. You cut down the trees and run cats all over the place and screw up all the old drainage patterns. And you don't replant. Then the heavy rains come in the winter, sixty or eighty inches of rain, and take the topsoil that's taken thousands of years to build up, and it runs down into the creek and screws up the fishing. The brush comes back—a lot of it. It becomes increasingly combustible. So some hot summer a fire comes along and burns it off. Then brush comes in to replace the hardwoods, a chaparral species like *Ceanothus* or manzanita. Then you get into a brush-burn cycle, and it's real hard to get a forest back in there. You get enough of that and change enough microclimates, then your macroclimate starts to change. It starts to dry up. The rain just runs off the soil, and you get this progressive drying out.

"About three or four years ago, I started daydreaming about taking this damaged environment and trying to put it back together. That was the first time I started thinking holistically about this place, about this watershed I live in. I thought about putting together a project that would take the out-of-work rural poor (which we have a lot of) and the damaged environment and put them together. A labor-intensive environmental repair.

"Ideally, I'd like to fix all nineteen square miles of Redwood Creek. I'd use controlled burning and a lot of planting. Live plantings, stakes, tan oak acorns, whatever works on a particular slope. To heal a watershed, everything is site-specific. You look at the slope. Are you really down in the subsoil? Is that all you've got? Is it wet in the winter? How wet? Does it dry out? Is it southern exposure or northern? All these things will vary. You figure out what will grow there based on local conditions—anything to get cover on the ground. Once you get that initial cover, then you can come back with a higher-level or higher-succession plant. If forest was there before, forest can be there again.

"One of the interesting things about holistic forestry is that when you look at the whole site, you find that a lot of the work fits together. For example, in timber stand improvement you generate a lot of waste material, and that can be used for erosion control, in contour wattles and check dams. You cure two problems at once. But most government programs and industrial forestry contracts talk about a contract for just one thing—a contract for timber stand improvement, or for conifer release, or for this or for that. And you get into some of the craziest goddamn situations that way. For example, suppose you are doing a timber stand improvement project; a gully runs through the forty-acre tract you are working on. You may be required by contract to burn the stuff you cut, instead of using it to stop erosion in the gully.

"Holistic forestry can be done—and it has been done right here in southern Humboldt. From any high point in southern Humboldt, you look out at the Tostens' ranch.[1] Their land sticks out like a green thumb. There were two differences. One, they owned the land and intended to stay there. Forestry was part of what they were doing, along with sheep ranching and all the rest. They looked at the land first and figured out what it could provide. The other thing they did was replant. It's so insanely simple, it almost drives you nuts. Now, their land is real healthy, real productive. You can see it, right down the borderline where the gyppos logged up one side and then split. But the Tostens intended to stay, and they did.

"The concept of holistic watershed repair is extremely logical. There's a natural, ecological, social, geographical unit. It's a whole piece, this watershed. When you work on it, it has favorable effects all the way down. The problem with it, of course, is there ain't no money for it. There's still a lot of money to be made exploiting land, and there's not a dime to be made in repairing it. It's enormously shortsighted. Can't we afford healthy land?

——— " ———

Forestry by Hand: Site-Specifics

The challenge of holistic forestry is to create alternatives to industrial monoculture that are economically viable. Unable to depend exclusively on government grants for environmental repair, holistic foresters must seek out techniques that can simultaneously heal the land and turn a profit. Is that an impossible task?

In fact, nature has its own healing mechanisms that might well be modified to suit human needs. Consider the properties of the red alder tree, for instance. Alders thrive on disturbed soil, and there is plenty of disturbed soil in the wake of most logging operations. As a pioneer species, alder is a quick-rooting tree that can penetrate compacted soil and loosen up the earth for successor species such as Douglas-fir. The alder roots serve to stabilize slumping soils. The leaves transpire excess moisture from the ground, and this, too, helps to prevent mass earth movements. The roots also host nitrogen-fixing nodules, and the leaves that the alders shed each fall turn into one of the richest of all tree litters in the United States.[2] As an added benefit, alders host certain mycorrhizae that minimize the incidence of root diseases, not only for themselves, but also for neighboring trees of different species. The red alder, in short, is an ideal healer of damaged land; it simultaneously penetrates, stabilizes, and enriches the soil, while providing a kind of inoculation against disease for other trees nearby.

Why, then, do industrial foresters treat the alder like a weed and try to eradicate it from their forested gardens? Because alders grow quickly, too quickly for the comfort of commercial tree farmers. Alders tend to dominate many sites, suppressing more valuable Douglas-firs. The industrial response is to eliminate the alders and "release" the conifers. The soil-doctoring trees are removed without being allowed to practice their medicine.

But what if the alders were left alone, permitting them to heal the soil? The conifers, according to this notion, would get their turn in due time. Alders might grow quickly, but they die quickly, too. After fifty or sixty years, the alders start to fall down and open the canopy for other types of trees. Gradually, more valuable timber will come to replace the alders. This system worked well for thousands of years, so why not allow it to continue?

The problem with this "natural" solution is that it ignores economic realities. Allowing prime commercial forest land to remain without saw-timber for half a century is not financially viable. As long as the issue is presented as alders versus conifers, the alders will continue to be merci-lessly eradicated.

Fortunately, there are other management alternatives that might be able to utilize the soil-healing properties of alder without sacrificing the greater commercial value of conifers. One system would allow alders to prosper for about ten or fifteen years in the wake of logging operations. During this time, the trees could be expected to grow to a diameter of four to five inches and a height of thirty-five or forty feet, accumulating a significant biomass that could then be harvested for use as energy. Interestingly enough, the alders' nitrogen-fixing capacity is greatest in the early years of growth and on severely depleted soils; after they have enriched the soil for about fifteen years, their nitrogen productivity tapers off.[3] By allowing the alders to thrive on their own terms for a rel-atively short period of time, foresters would reap most of the benefits from their soil-healing properties. And they would reap economic ben-efits as well. If we assume a significant increase in the demand for wood as fuel in the near future, the young alders could pay their way out of the woods, and the land would never really be out of production.

Tree farmers who utilize alder in this way would be planting a "green manure" crop, just as traditional farmers have done for years. There is nothing really new or radical about the idea of rotating crops to save the soil from eventual exhaustion. In the long run, biological fertilizers are much more effective than chemical fertilizers, which produce no "carry over" effect beyond their immediate application.[4] Indeed, it is strange that the timber industry, committed as it is to the agricultural model, has not yet embraced the idea of crop rotation. If alders are allowed to dominate a site for 50 years, the resistance to "green manure" is under-standable; but is 15 years too long to wait for healthier soil?

This is not, however, an ideal solution. Although crop rotation might be preferable to current management practices, it still would be a form of even-aged monoculture; at any given point in time, there would be only one dominant species on the site. How would alder be kept in check after its turn was finished? The temptation to use herbicides would be just as strong as ever.

Red alder: weed tree or natural fertilizer?

Holistic foresters, with their preference for mixed rather than pure stands, would prefer to allow commingling between alder and Douglas-fir. Back in the winter of 1932–33, red alder seedlings were interplanted with four-year-old Douglas-fir in the Wind River Experimental Forest in southwestern Washington. At age 30, the Douglas-fir in the mixed stand was equal in volume to the Douglas-fir in a control plot; by age 47, the volume of the Douglas-fir in the mixed stand was significantly higher.[5] Including the volume of the alder, twice as much wood was produced in the mixed stand—and the soil which would grow the subsequent generation of trees was much richer. The alders, while nourishing their neighbors, had achieved merchantable size themselves—two crops for the price of one.[6]

Unfortunately, the intermingling of alder with Douglas-fir is difficult to implement. Since the alders grow far more quickly than the conifers during the first few years, they would have to be held in check, yet not totally eradicated. Area-wide treatments, such as the aerial application of herbicides, might not be able to maintain a healthy balance between the species. Manual techniques, on the other hand, would have a better chance of achieving an appropriate distribution. Laborers would move through the forest, cutting down a few trees here and a little brush there in order to obtain the desired result: a genuinely mixed stand. Choices would be made on-the-spot by individual workers, not by managers in distant offices. This is labor-intensive, site-specific forestry; according to some holistic foresters, it is the only real alternative to the mechanized, computerized techniques that now prevail in the timber industry.

Rick Koven

Member of Great Notions, a Reforestation Cooperative

"A couple of years ago, some people in Oakridge wanted to have a 'brush-in.' The government planned to spray herbicides on some acreage that was right on these people's water source in Salmon Creek. There was a lot of controversy about it. So a bunch of people went up there and cleared the brush by hand right before it was to be sprayed by helicopters. None of these people were cutters. They went up there with machetes and kids and everything. It took them two days to do twenty acres. Apparently they did a terrible job, but it was politically effective.

"So the next week, some environmentalists down in the Cottage Grove district wanted to do the same thing. We called them up and told them we'd go down there as a crew and cut the unit with saws. It would still be a 'brush-in,' but this time we wanted to do it as a professional crew and not just as citizens. It worked. We went through thirty-five acres in about two hours. It immediately dawned on us that there was a lot of money to be made in brush clearing.

"That summer the Forest Service started letting out contracts. We got our first one for 130 acres, then we got another one at the end of the summer, one in the Coast Range and one in the Cascades. It worked out really well. The Forest Service and BLM let out 3,500 acres all together. They had competitive bidding on the open market, and, as far as I could tell, it was a really positive thing.

"Manual release is a silvicultural issue as well as an environmental issue. Let's say we woke up tomorrow and the EPA had just proved conclusively that herbicides were absolutely safe. Would that mean you should never use manual release? I say no. Why should you use manual release, when it costs at least somewhat more than herbicides? The biggest reason is because it's site-specific forestry. At the exact time and place of application, you can make a specific decision about what is best for that site. When they take a unit and spray it, there's no decision. The decision is based on a computer, saying this plantation is seven years old, and looking at the photograph it's got some brush on it, so you spray the whole thing.

"You can get into self-defeating practices that way. For example, I was in a unit the other day that was sprayed last year. But there was never really much brush there. We could have manually released it for $25 an acre. And what was the result of spraying? For some reason this area had tons of cedar reproduction. Just tons of it— I've never seen anything like it. But all the cedars were dead. Now if I could have gone in there and manually released it, I could have saved the cedars. A lot of them were right on the road banks. They were tight little clumps, which is great for holding erosion on cut banks. Now they're all dead. So what was the result of spraying? They hurt themselves. Eventually they're going to have clogged up culverts from erosion. They're not going to have any cedars up there, and cedar's a valuable tree. So what's the value of spraying? It's so arbitrary, and sometimes it just doesn't make any sense.

"With manual release we come in and say, 'What does each microsite need?' We make a human decision. Let's say I'm in a slide, and there's lots of erosion going on. I look in there and I see all these alders and brush, and they're holding that slide together. There's a couple of little suppressed firs. I say, 'If I cut those alders, that soil comes down and there's no fir.' So I don't cut them. I tell the inspector, 'Look, I didn't cut those trees because you need them there.' That type of situation repeats itself over and over in different sets of circumstances, where the worker can make a decision based on a microsite observation at the exact time of application. That's very important. That's good forestry.

"In site-specific forestry, instead of releasing every tree in the unit, you release the trees that you decide will be most commercially valuable. Therefore you don't use strict spacing, you go for reasonable spacing. You choose the healthiest trees. Don't say, 'We have 1,000 acres, and we want 520 trees an acre, and therefore the spacing is ten-by-ten, and therefore you release every tree ten-by-ten. . . .' That's how the big companies work. It's strictly geometrical.

"In site-specific forestry, you're on a south slope, 45°. The main reason to release is sunlight, so why should you release the north side of the tree more than a foot? There's absolutely no way you can shade the tree from the north side. On the other hand, the more you release downhill, the more sun you get in the early morning and late afternoon. If we ask the silviculturalist to put something like that in our contract, he just says, 'We have a problem if we put anything in our contracts that's a little complicated, because the workers don't understand it.' That's bullshit. Everybody knows where the sun is, what the directions are. Everybody knows where north is. But they want to have a uniform standard and plug the worker into it.

"The workers can think, but they don't trust the worker to think. Give the workers any leeway and they'll do the easiest thing. The workers are stupid and lazy, right? Wrong. Forestry has changed. The workers aren't as transient as they used to be. You've got a lot of college education out there. You've got a lot of interest in forestry. That's a big change. And the old guard in the Forest Service doesn't like that. They can't deal with you the way they dealt with the bums. You question, you demand, you're a thinking human being. They're not ready for it. They have an image that the tree planter lives in cheap hotels, goes on long drunks between contracts, and cashes his paycheck at the bar. Whatever you tell him to do, he does.

"This leads into the labor issue. What's happening is that they're trying to reduce the number of people working in the woods. It's automation, basically. But what *we* think is that good forestry is people going out and making decisions on a real small scale. What *they* think, apparently, is that good forestry is very few people making broad decisions based on vicarious or indirect observations—aerial photography, computer projections, and so on. You get the impression that they would like each district office to have one forester, a room full of computers, and twenty secretaries to do all the bureaucratic paperwork. The forester just sits there and presses buttons, and that's supposed to be forestry.

"What we're saying is that forestry is thousands of people working in the woods and making money. That's good for the economy, and that's good for the forest. You have people out there looking, seeing, thinking, talking, communicating. Maybe they don't want people to see what's going on. That's very basic to the Forest Service and industrial foresters' opposition to this whole movement. They're very strong on the division of labor. They don't like to see workers integrated into decision-making positions. There's something about that that bothers them. But workers *can* do research and make meaningful decisions. We're trying to develop a relationship between the worker laborer and the scientist or technologist. It should all be integrated. That way you have a built-in interest in your work.

"You've got to have some interest in your work just to survive. Otherwise it'll wear you down, because it's not easy work. You get on the job at dawn, really early. You get out and you pack a gallon of gas and oil. You use a good, light saw. A heavy saw would kill you in this work, because you have to cut twenty to fifty thousand stems an acre. Your saw is always on full. You go down to the unit and you work six hours straight and then leave—no lunch or anything.

"All the units are so different. In one case you're surrounded by alder whips, thousands of them. You bend over and just mow them down. Then you encounter a wall of salmonberry; blackberry trailers are around your neck. You start reaching back with your saw to release yourself. It's like Vietnam. You get all wrapped up in brush and you try to cut yourself out. The Coast Range is tough to work in because it's so steep, so incised. Every unit is just ravine after ravine, so the surface area of the unit is twice as much as the actual linear acreage. But it's spotty. A lot of these units include parts where there's nothing to do. That's when you make your money.

"It's not easy work. And it's dangerous because you're cutting off all those little, short sticks, and when you fall down on them they stab you. And you're always getting whipped in the face and strangled by the brush. But it's not really serious. Mostly you get slapped in the eye or whipped in the lips and the nose. If you shut off your saw and start listening to your friends, you hear guys cursing and screaming. But you know they're not hurt. You know they're just getting whipped.

"It's not like tree planting. You can't talk when you work. It's too loud. You wear earplugs. If you want to take a break or talk to everybody, you have to wait until their saws idle down, and then rev your saw five times and maybe they can hear you. Then everybody takes their earplugs out.

"It takes a very special person to do this for more than five to ten years. It's worn my body down physically, very definitely. Real chronic problems: my back, wrists and elbows, shoulders. You find that with a lot of tree planters. Bad backs from bending down all day, tendonitis in the wrists and elbows from the shock of hitting your arms on the hard ground. For brush cutting, the most common thing is 'white finger' from the vibration of the saw. Like

I said, it's not always pleasant. You're out there working in the cold rain. Your joints ache. You resent the government people who sit around in an office drinking coffee all day.

"I don't want to do this work forever. I don't like it that much. The best part of it for me is when I'm on that bus leaving the job. On the other hand, when you're in the age group of eighteen to thirty there are a lot of unemployed people. It's good experience; I've learned a lot from it. I've learned a lot from being in a co-op; I've learned a lot from being in the woods; I've learned a lot about myself physically. You love it and you hate it. It's what we do. If I don't want to do it anymore, I'll change that. But right now that's what I do.

———— " ————

The Politics of Brush

Manual release is touted as an alternative to herbicides, which are alleged to present health hazards to both people and forest wildlife. Throughout the 1970s, phenoxy herbicides such as 2,4-D and 2,4,5-T were sprayed routinely on commercial forests in order to kill off or retard broadleaf plants. When mixed together, these two substances became the infamous "Agent Orange," which was used to defoliate the jungles of Vietnam. How safe was Agent Orange? According to many Vietnam veterans, it led to a wide variety of human illnesses. Several different laboratory tests linked phenoxy herbicides with embryo and birth defects in chicks, mice, rats, and hamsters.[7] The herbicide 2,4,5-T contains small amounts of dioxin, which an Environmental Protection Agency spokesman once described as "perhaps the most toxic small molecule known to man."[8] Laboratory monkeys exposed to minute quantities of dioxin perished in a matter of weeks.[9]

During the late 1970s, increasing numbers of residents in forested regions reported that they or their animals became ill when nearby areas were sprayed. In response to these reports, Dr. Ectyl Blair of Dow Chemical continued to maintain that the herbicides were perfectly safe: "We think the hazards of 2,4,5-T are really quite small. We've had it in agriculture for thirty years or more and don't believe that the spraying of 2,4,5-T is in any way related to the illness reports."[10] For many years, the Environmental Protection Agency agreed with Dow Chemical, sanctioning the use of phenoxy herbicides on forested land. When the evidence of ill effects continued to mount, however, the EPA finally outlawed any substance containing dioxin.

Herbicide spraying by choppers: how lethal is the dose?

Today, industrial foresters have switched over to a new chemical: triclopyr, which Dow uses in a product called Garlon. Although it breaks down faster and contains no dioxin, triclopyr closely resembles 2,4,5-T, the only difference being the substitution of a nitrogen atom for a carbon atom. Like the phenoxy herbicides, triclopyr promotes an uncontrolled expansion and division of cells in broadleaf plants. Although it currently has the official blessing of the EPA, environmentalists remain skeptical. Registration with the EPA, they claim, does not imply that a chemical has been tested sufficiently to establish its safety. Wasn't 2,4,5-T officially sanctioned for many years—until the evidence against it proved overwhelming? Why does the practice of forestry have to involve the purposive poisoning of trees, the inadvertent poisoning of wildlife, and the possible poisoning of people? Can't we find a better way?

According to its advocates, manual release is superior to the use of herbicides in four important respects: (1) It avoids the use of toxic substances; (2) it is a more flexible tool than herbicide application, since each site gets treated individually; (3) it retards the growth of competing vegetation without killing it off entirely (unlike herbicides, it does not simplify the ecosystem); and (4) it provides meaningful and timely employment. On the Hoopa Indian Reservation in northern California, for instance, the Tribal Council voted to ban the use of herbicides in 1978—not just for health reasons, but for cultural and spiritual reasons as well. Yet to remedy the poor forestry practices of the past, some form of vegetative treatment was necessary. Since 1979, hand release has been practiced frequently and successfully. In the long run, the improved forests will lead to more money for the Hoopa people; in the short run, the manual release contracts create work for Native Americans in a region where there are few other job possibilities. Since release work can be performed year-round, local loggers are able to find employment in the off-season.[11]

Critics of manual release observe that repeat treatments are often necessary, since the brush is cut back but not killed. They also observe that it is difficult and dangerous work. Why do the job by hand, they wonder, when machines and chemicals can do it for you? Why turn back the clock of progress? One text on vegetation management notes that manual release is a "highly hazardous occupational practice" because it "involves power saws and machetes." The aerial application of herbicides, on the other hand, is portrayed as "the safest, most efficient, and most cost-effective" method of brush control.[12] Upset by the portrayal of herbicides as a health hazard, the advocates of chemical treatment try to utilize the "safety" argument for their own purposes.

The strongest argument against manual release is its cost. Although bid prices range widely, it does appear to be more expensive than the aerial application of herbicides.[13] But how appropriate are straight comparisons between the two techniques? Politically, the use of herbicides is not always a viable option—and even when it is, there may be significant but hidden legal costs. Other variables also skew the results of any comparison, for they are used in very different contexts. When a group of researchers examined over sixty units slated for herbicide spraying within the Willamette National Forest in the late 1970s, they found that convenience rather than necessity served as the most important criterion for which units would be sprayed. All but four of the units were within one mile of a helipad, while the remaining four were within two miles of a helipad. Since the economics of helicopter spraying involves high expenditures for fuel, equipment, and chemical handling, large acreages have to be sprayed in order to lower the per-acre costs. Consequently, every unit that appeared moderately brushy and was within striking distance of a helipad was treated. In some cases, the units were too good to be sprayed: the crop trees were already topping out the brush. In other cases, the units were too bad to be sprayed: even if the brush were set back, there was not enough conifer stocking underneath to make any difference. Meanwhile, other units in need of management yet not proximate to a helipad were neglected.[14] Although the per acre cost of spraying herbicides is certainly less than that of manual release, do the savings make up for the fact that many sites fail to receive the precise treatment they need?

In some minds, the mere presence of brush seems to constitute a sufficient justification for chemical treatment, whether or not the brush is significantly impairing the growth of crop trees. The very word *brush* is pejorative. Dow Chemical, attempting to capitalize on this negative image, advertises its herbicides by depicting dragon-like monsters hovering over the woods; these dragons, according to Dow, represent "The Brush Demon" which must be "tamed" with chemicals.

Ecologists, on the other hand, prefer to use the friendlier term *pioneer* instead of *brush*. Pioneers, they say, are important components in a successional chain, not wicked weeds. The various species of *Ceanothus*, for example, are particularly troublesome to work with manually, but they are also of particular benefit to conifers. Not only do they fix nitrogen and protect against browsing animals, but they also create an environment less favorable to insect pests and root pathogens.[15]

Vegetative management is among the most controversial issues in forestry. Traditional foresters have focused on the notion of *competition* for available sunlight, water, and nutrients. Increasingly, many foresters are looking at ecosystems in terms more suggestive of *community* or *cooperation*. This difference of outlook resembles the classic ideological division in politics between the "right" and the "left." If we focus on individual survival, we see the forest as a kind of capitalistic society in which each organism tries to out-perform its neighbors; if we focus instead on the good of the community as a whole, the forest ecosystem resembles a socialistic society in which the well-being of separate individuals is achieved collectively through a complex network of symbiotic relationships.

This difference in ideology permeates the field of forestry, affecting not only the obvious political controversies between the timber industry and environmentalists but also the focus of scientific research. Depending on their outlook, different scientists will document events that reveal the importance of either competition or community. In 1990, three scientists studying a 9-year-old plantation of Douglas-fir concluded that the diameters of the crop trees were only half as large in areas of high madrone density than in areas where madrone was entirely absent.[16] In 1989, two scientists demonstrated that the basal area growth of Douglas-fir seedlings could be doubled by transferring small amounts of soil from a madrone stand to the planting holes on certain sites.[17] Are these studies contradictory? Not at all; they simply examined different variables. The first study offered an example of the adverse impact of competition; the second study showed that one species can provide a hidden benefit to another.

As in human society, "competition" and "community" are both relevant concepts in forestry. Although ideologues tend to focus more on one than the other, a holistic forester must see that both competition and community are omnipresent within the forest. Management treatments, of course, will vary according to the specific time and place. When madrones threaten to take over a site, they will have to be held in check (although not necessarily eradicated) in order to allow for conifer production. When conifers are hard to regenerate, on the other hand, the seedlings might well benefit from the presence of some madrones. The optimal number of madrones in a forest can only be determined by a site-specific analysis, not by some *a priori* belief.

Recently, the issue of vegetative management in northern California has taken a new turn due to the work of holistic forester Jan Iris. Instead of just worrying about how to check the growth of hardwood "weeds," Jan had another idea: Why not put them to good use?

Peggy Iris

Wild Iris Forestry

"For Jan, forestry was very much like gardening. Jan loved roses as much as he loved trees, and there really wasn't very much difference between what he was doing in his garden and what he was doing in the woods. You release a plant, you weed around it, you clean it up, you bring it into a healthier condition so it can grow strong, whether it's a tree or it's a rose bush.

"About twenty years ago, Jan fought the Finley Creek fire. He had never really experienced anything like that before, a real forest fire covering thousands of acres. Living in the deep woods as we do, with the state of the forest being as unhealthy as it is, so brushy and overcrowded, it didn't take much of an imagination to see what a fire like that could do to your homestead. The lasting impression of that Finley Creek fire inspired a lot of the fire hazard reduction that Wild Iris started doing in the woods.

"We started by using some of the cost-sharing money from the federal FIP (Forestry Incentives Program). It was actually designed for conifer release, not fire hazard reduction, but since there were some conifers on the land we worked, we qualified for that program. Jan started by limbing the trees as far as he could reach with a chain saw, breaking up the fire ladder. He thinned out some of the hardwoods to reduce the fuel, but then he started asking: 'What are we going to do with this wood?'

"He started to see that fire hazard reduction fit into the same model as stand improvement. The forest was so overcrowded after the logging of the '40s and '50s. Just reducing the fire hazard, thinning the forest, allows the trees to grow the way they should. It accelerates the growth and improves the quality of the wood. The limbing eventually creates clear lumber, and of course the trees grow straighter. In terms of long-term management, you're creating future wood supply. The healthy forest creates the highest quality lumber.

"Jan defined holistic forestry as all-age, all-species management; all species referring to not just the trees but all of the fauna and flora down to the mycorrhizal fungi. He wanted to restore the forest to predisturbance composition in terms of the mixture of tree species. But that's pretty tricky. What do we mean by 'predisturbance'? How far back are we going to go? What about the changes of composition through the millennia? I don't think anybody has a very clear idea about how that all works. We know obviously that there were a lot more conifers and a lot less hardwoods before the white man started logging, so we wanted to get back to that mix. But we didn't want to eliminate the hardwoods altogether, like industrial forestry does.

"When Jan started out, just about everybody regarded the hardwoods as total weed species. There were a few people locally who were playing with milling hardwood and air-drying it, but they were finding out that the wood warped and twisted and checked. It was pretty hard to cure. There were a few experts who realized that it was technically possible to make lumber out of the native hardwoods, but they too advised us against it. They said that the reputation was so bad that even if we were able to produce lumber, we wouldn't be able to market it at all. People would *never* buy tan oak, no matter what we did with it.

"Jan wasn't convinced. He went down to the U. C. Berkeley forest products lab and took a wood-drying class. Each evening after class, he would go down to the archives in the library and dig around to see what he could find on native hardwoods. There was very little, but he uncovered this information about Mal Coombs' kiln and the work that he had been doing in the '50s. Jan documented that there are floors and paneling throughout Humboldt

County from the tan oak and madrone that Coombs put out. But Mal couldn't make it economically because he was competing against red oak from the East, which was very plentiful, high quality, and cheap. The problem was that our native hardwoods took so much time to cure—at least a year—and it's economically a hard thing to sit on wood for that long. So Mal's kiln is just sitting there. Nobody took it up after he died. Everybody just thought he was crazy.

"Jan came back from Berkeley very excited with all this information about how to process the wood. He had worked out a kiln design with the professor down there which was specific to drying our native hardwoods. We invested all of Jan's and my savings into the kiln, and we went to work.

"We milled on site in the woods, then stacked it and stickered it and air-dried it in the shade. It was best to cut in the late fall or early winter, so the initial drying would be very, very slow. We thought at that point that it needed to air-dry for a year, but now we've found that we can usually put it into the kiln just before the real heat of the following summer. It stays in the kiln a week to ten days. When it comes out of the kiln, each board needs to be looked at and evaluated with care, because that's when we need to determine the highest use: what boards are furniture quality and what boards should be turned into flooring or paneling or whatever.

"The kind of forestry which Jan practiced was very time consuming—not only the time to process the wood, but the extra time he took at every stage. He spent many hours looking up into the canopy and considering how much light he was allowing into the forest. He also had a very inefficient yarding system because he was so careful not to do damage to the forest floor or to the other trees. He hung lots of blocks as high as he could get them, and he would often have to redo the blocks and zig-zag in some crazy pattern in order to avoid the standing trees. It was taking forever, and that means that it was expensive. We soon realized that the only way that we would ever be able to pay for this kind of work was to add value to each piece of wood that we cut; selling green logs would not even begin to pay for the operation in the woods.

"As we started to create a market for hardwoods, we also started to worry about the new hardwood industry repeating the ecological degradation of the softwood industry. The hardwoods are turning 180 degrees around, from having been considered a weed species to being a resource that's very sought after. Are we creating another boom that will just lead to another bust? It's pretty frightening that we might have opened a door that could lead to another wave of environmental degradation. How can you be sure that the hardwoods you buy in the lumber yard aren't from some clearcut or from an area which has been sprayed with herbicides?

"In the spring of '90, Jan came to me with this idea about starting a certification and labeling program for ecologically harvested forest products, based on the model of organic food certification. I got very excited about it. People were starting to call us from all over the country, wanting wood that they knew had been holistically harvested. Controversy over tropical rainforest issues had made woodworkers look more toward native species. People everywhere were really starting to care where wood was coming from.

"The problem with certification is that it's very easy for the industry or anyone else to claim that their wood has been sustainably harvested, particularly if they have a lot of money behind them. People see a full-page ad, and they really want to believe it. This is called 'greenwashing': using the terminology of sustainable forestry for nonsustainable products. Anyone can claim anything. It's like saying something is 'all natural' in the food industry.

"The only way to fight greenwashing is to set criteria for what is actually meant by 'sustainably harvested,' and then to track and label that wood. This would be a wonderful service to the consumer, and it would also be a way of charging more money which could trickle back down and help recoup the expenses of doing a slower, more careful style of forestry. We're now at the point where we are about to certify wood and other products with a label of PCEFP—Pacific Certified Ecological Forest Products.

"About five years into the development of Wild Iris, we became aware that Jan had prostate cancer. We were able to get him into remission for about a year, using experimental hormone drugs. I worked very closely with him during that time, getting a quick course in the kind of forestry he practiced, the processing of the wood, and how to run the business. During that year we came up with the idea of forming a nonprofit organization. We had been spending a lot of time on education, traveling around and sharing our ideas. We also saw the need to get more scientific. People had told us we should document our work with inventories and hard data. That was part of the motivation to create the Institute for Sustainable Forestry. Also, since Jan had cancer, I was very much aware that we had to come up with some kind of a scheme which would stabilize the concepts. It was an incredibly scary thought to me that I might be the one, individual person who was supposed to carry this model forward if he were to die.

"Shortly afterwards, Jan started getting really sick. The cancer broke through the drugs, and we realized it would be a real uphill battle to keep him alive for any length of time. That was when Jan received the award for the 'Stewardship Forester of the Year' from CDF [California Department of Forestry]. It was the first year the award was ever given. The decision to give it to Jan was very controversial because he was not a professional forester, a legal RPF [Registered Professional Forester], and there were people on the committee who were very offended by the fact that someone like Jan, who had a very limited educational background in forestry, should receive this award. They felt like: 'Who is this guy, anyway?' But he still got the award because his work was so respected.

"Jan died two years ago, but his work is being carried on by Wild Iris and the Institute for Sustainable Forestry. The Institute is doing the kind of documentation which Jan had always thought would be necessary. In our pilot watershed project, we're monitoring the cost and the productivity of everything that happens, from the management plan all the way to the marketing of the wood. We're looking at the recoverability rates of different species of trees, how much of what grades of wood are coming out of the different ages of trees, the composition of the stand, everything. It's quite complex. We're trying to keep track of all that and get it on the computer, including all the costs. Then we're looking at the market: who it is, and what the market will bear in terms of increased prices for certified wood.

"We need to learn as we go along what's really happening, to monitor closely. We need to have a sense of humility in terms of ecosystem management, and to realize that we might be wrong. We have to do this in a slow enough way and a gentle enough way so that we can monitor it and make adjustments to what we're doing before we do some serious damage. We can't claim that we know all the answers, because we don't know exactly what's happening in terms of the interrelatedness between all the species, all the fungi and the mycorrhizae. Nobody knows. The research just isn't there.

"It's a whole other way of doing forestry. We're not looking at a quarterly profit; we're looking at a very long-term sustainability through many generations of people and trees. The real brilliance of Jan's vision was tying the economic structure of community-based forestry into the definition of sustainable forestry. The economic part of it fits completely in with the ecological side. One doesn't happen without the other. You can't expect to have a stable economic base without ecological forestry, and vice versa. The two go hand-in-hand. The economic structure of the area is based on the stability of that resource. And the best way to insure sustainability is to have the decisions being made by the local people who have a real interest in the future of the forest. If we want to break the boom-and-bust cycle which is so common in rural areas such as ours, if we want to create jobs for the young people of the community, the people living in an area have to maintain control of their resource base.

———— ” ————

Holistic forestry, in that it deals with whole systems, must include all relevant variables. Artificial dichotomies—economics vs. ecology, competition vs. community, man vs. nature—must be overcome. In this day and age, we cannot realistically deny that man is a part of nature, that human economy and forest ecology are interdependent. Significantly, holistic forestry differs from preservationism in that it is an active, not a passive, approach. Holistic forestry acknowledges the economic argument, affirming our right and need to employ people and to use the forest as a resource, while simultaneously demanding that we treat the forest respectfully if we are to continue to coexist.

The fundamental principles of holistic forestry are as simple as Zen precepts: use manual labor whenever possible; avoid artificial substances; treat the specific needs of each site; pay attention to the interconnections between sites; find strength in balance and diversity; and—the golden rule—complement, rather than contradict, nature's actions as much as possible.

Holistic forestry is still in its infancy. Although its ideas are gaining wider acceptance, there are only a handful of "practitioners" out in the field. Holistic forestry postulates a middle road between overmanagement and no management at all, but the exact route which that middle road will take has not been determined. The basic ideas seem sound. But where do we go from there? How do we implement these ideas in the real world?

5

SILVICULTURAL SYSTEMS

The most fundamental decision in forestry—a decision that affects all subsequent activities—is which silvicultural system to use. Since the term *silviculture* means the cultivation of forest trees, one might suspect that silvicultural systems would be classified according to regeneration strategies. Technically, this is true, but in practice the methods of regeneration are strongly influenced by the type of harvesting that has just occurred. Hence, the various silvicultural systems take on the names of their corresponding harvesting components.

There are three basic types of silvicultural systems in common use today: clearcutting, selective cutting, and shelterwood cutting. In *clearcutting*, all the trees are harvested at once, creating a large patch of open ground in which seedlings can be planted. In *selective cutting*, the trees to be harvested are chosen individually or in small groups; a major portion of the forest is left undisturbed, providing a continuous cover for the land and a source of seeds for the next generation of trees. The *shelterwood method* is a sort of compromise between clearcutting and selective logging: most of the trees are harvested at one time, but enough healthy, mature trees are left standing to provide a natural seed source and a modest amount of shade for the next generation. Once the new seedlings are firmly established and no longer require a protective cover, the sheltering trees can be removed.

The proponents of clearcutting feel that history is on their side: long before modern logging, huge wildfires periodically devastated the forests to create nature's own clearcuts.

The proponents of the selection method feel that history is on their side: long before modern logging, nature practiced selective cutting by culling out the weak, diseased, or aged trees and opening up small patches of light within the otherwise shaded forest.

The proponents of the shelterwood method feel that history is on their side: long before modern logging, small ground fires swept through the forests and removed everything but the hardy, mature trees that towered above the level of the flames. It was nature's way of opening up the brush and preparing the ground to receive the seeds from the tall and majestic survivors.

Which silvicultural system is really the most natural? Which does the least environmental damage? Which is the most efficient? Which is best suited to provide a vigorous start for the upcoming generation of trees?

Clearcut Controversy

Most of the arguments for and against clearcutting were presented in Chapters 2 and 3, but I'll summarize them here. Clearcutting offers unparalleled opportunities for centralization and efficiency. All harvesting occurs at a single time and place. Machinery can move around freely without fear of damage to residual trees; once the machinery is brought to the job, it is allowed to produce to its maximum capacity without external constraints. Clearcutting is logging pure and simple. Once the trees are logged, the area becomes more accessible to other management techniques, such as burning, replanting, and spraying. Clearcutting creates a blank slate upon which the modern tree farmer can create a new forest according to his own notions.

The most popular argument for clearcutting among commercial foresters stems from a classification of trees according to their tolerance for shade. Young seedlings of tolerant species prosper under the shade of an existing forest canopy, whereas seedlings of intolerant species require direct sunlight to produce satisfactory growth and development. Many commercial species—including Douglas-fir, the nation's most important source of lumber—are classified as intolerant. These trees need the sun, and clearcutting gives them all the sun they could ever want. Clearcutting, say some foresters, is not just desired for its efficiency, it is actually required for the satisfactory reproduction of intolerant species.

But does the entire forest have to be cut down for the sun to reach the ground? If only a handful of trees are removed, the sunlight can still infiltrate the vegetative cover. Group selection— where the trees are cut down in small patches throughout the forest—satisfies the biological criteria for intolerant species by allowing the sun to shine on young seedlings. Since alternatives exist, clearcutting cannot be considered a biological necessity.

Clearcut patchwork in the Northwest

Nor is clearcutting the functional equivalent of a wildfire. Fire might kill the trees, but it does not immediately remove them. An area burned by wildfire is covered with snags, and with the insect-eating birds that live in them. The ground remains partially shaded from the sun and protected from the elements by these snags and by scorched foliage. The soil in the wake of a wildfire is not compacted or disturbed by heavy equipment. The nutrients lost due to wildfire will soon be replenished by pioneer brush species, the very plants that are seen as enemies by contemporary industrial foresters. Nature's response to wildfire is to cover the ground with vegetation as quickly as possible; industry's response to a clearcut is to keep the ground open and free of "weeds." Wildfire is a part of a successional process; clearcutting is part of an attempt to bypass that process.

Clearcutting might be economically efficient, but the attempts to justify it with ecological arguments fall short. The environmental drawbacks to clearcutting are real, not imagined. When the vegetative cover is entirely removed, ambient forest temperatures become more extreme, hotter in the summer and colder in the winter. The ground is wetter and less stable during the rainy season, but becomes hard and crusty during the dry months of the year. When totally exposed, the soil is subjected to splash and sheet erosion during heavy rainfall. Soil nutrients are depleted when the entire forest biomass is carried off on a logging truck or burned in the aftermath of a clearcut.

Since natural seed sources are usually destroyed, the area must be artificially seeded or planted. Man's chosen seedlings will be forced to cope with the increased environmental extremes and will face heavy competition from pioneer, sun-loving species formerly held in check by the shade of the forest. To overcome the competition, the commercial trees will often require the aid of human benefactors: herbicides will be sprayed over the new plantation, subjecting animal and human life to unknown hazards. In terms of the overall health of the forest environment, clearcutting, except in certain special circumstances, has serious drawbacks.

Getting Selective

Is selective cutting more natural than clearcutting? When nature culls out individual trees from the forest, she lets them rot on the ground; she does not build logging roads throughout the woods to remove the fallen carcasses. In selective logging, roads are built anywhere and everywhere. It is difficult (although not impossible) to log selectively with roadless methods, such as skyline or balloon logging. In order to obtain as much timber as could be produced in a 40-acre clearcut, perhaps 100 acres or more would have to be logged selectively. To reach the scattered trees, a maze of haul roads and skid trails would have to be constructed.

Roads, we have seen, are the major cause of landslides and stream sedimentation. How can selective logging be environmentally flawless when it requires more roads than a corresponding clearcut? The area covered by road rights-of-way is forever lost to production, for the roads are used repeatedly to cull the trees. In a selectively managed forest, the timber is harvested as it matures, perhaps every ten or twenty years. The roadbeds themselves never have time to produce a new crop.

In a mature forest, trees tend to protect each other from the sun and the wind. When some but not all of the trees are removed, those that are left are exposed and vulnerable. They are more likely to blow over in a storm. Their lower needles are sometimes scorched by direct sunlight, and even their trunks can be streaked with sunscald. The repeated entry of heavy equipment into the forest creates serious problems for the trees that are left standing. The residual timber is often delimbed by its falling neighbors, and the trunks are sometimes scraped by cats or cables. The damaged trees are, as expected, more susceptible to various diseases.

In addition, the frequent use of equipment compacts the soil, causing the residual trees to grow less rapidly. In an experiment conducted on loblolly pines, a tractor pulling a load of logs was driven past some trees six times to simulate a normal logging operation. The experiment took place in wet weather, maximizing the extent of soil compaction. Five years later, the growth rates of the affected trees were measured and compared with the growth rates of a control group. Where the tractor had passed on one side of a tree, there was only a slight decrease in the growth rate. But where the tractor had passed on two sides, the decrease was 13.7%; on three sides it was 36.3%; where the tractor had passed by on all four sides, the trees had lost 43.4% of their expected growth.[1]

The damage to residual trees is subject to some control. If the trees are selected well and logged carefully, negative effects can be minimized; but there is still the problem of regeneration. How can the seedlings of shade-intolerant species be appropriately nurtured if the forest canopy is not removed? If only one or two trees are logged from any given location (individual selection), the growth of the seedlings might be relatively slow. If several trees are taken from the same area (group selection), the sun will reach the ground and regeneration can proceed apace. But where do we draw the line between group selection and a clearcut? How many trees have to be taken before the ground begins to suffer from overexposure and other problems created by clearcutting? As selective cutting moves toward clearcutting by increasing the number of adjacent trees that are logged, the problems of one are replaced by the problems of the other.

Historically, the greatest problem associated with selective cutting has been *high-grading,* the taking of only the most desirable trees. As long as there was old-growth timber for the asking, loggers saw no reason to take anything but the best. But taking the best means leaving the worst, and that's exactly what they did. Residual trees were left to re-create subsequent generations, but these parent trees had been inadvertently selected by a sort of reverse genetic engineering. Today, we have inherited the results of this high-grade selective logging: throughout the backcountry we are left with low-quality residual timber and natural regeneration from inferior parent stock.

Selective logging, of course, does not necessarily entail high-grading. If the harvested trees are chosen correctly, the quality of the future forest can be improved rather than degraded. Indeed, a well-managed selective system has many positive attributes. When seedlings are nurtured in the partial shade of a protective forest canopy, they tend to grow upward instead of outward. They do not put their energy into the sizable limbs that characterize trees which grow in the open, and their wood is consequently less knotty. The same shade that tempers their growth increases the quality of the lumber which they will eventually produce. The grain is relatively straight and fine, and the wood fiber is actually stronger. Commercially, this means that a selectively managed stand of timber can actually yield greater financial returns than a comparable stand grown in the wake of a clearcut.

There tends to be a greater variety—and therefore greater security—in a selectively managed forest. The trees are of different ages, at different stages in their growth cycles. They root in different soil strata and utilize different proportions of nutrients. The accompanying vegetation is generally more diverse in an all-age stand: a shade-loving understory; tolerant species of trees amidst the crop trees; and, a sprinkling of pioneer brush species in the partial clearings.

With greater natural balance, the ecosystem requires less tinkering to keep it in line. Since the danger of insect and disease epidemics is lessened, fewer insecticides and pesticides will have to be used. Since there is no point at which brush threatens to take over the entire forest, there is less dependency on herbicides. Since seeding will often take place naturally from local, well-adapted parents, there is less of a need for elaborate nursery and seed orchard programs.

Finally, selective management systems tend to protect the soil from the ravages of overexposure. There is less nutrient loss due to leaching, less sheet erosion, and less runoff during storms. Since a partial cover of trees is left after each logging operation, there are always roots to hold the soil intact, and leaves or needles to transpire excess water; the ground is, therefore, less likely to collapse from saturation. Although the soil as a whole is not as wet, the upper layer of earth can actually retain moisture better after a selective cut than after a clearcut. Since there is still a vegetative cover, and since a layer of litter still protects the earth, there is less surface evaporation from direct exposure to the wind and the sun. During the dry summer months, the ground tends to be damper and cooler under a forest canopy than out in the open. Seedlings are less likely to die of thirst, and a fire is less likely to break out. And, since moisture and temperature levels are more constant in a selectively managed forest, the soil environment remains hospitable to friendly fungi, bacteria, worms, and other organisms that decompose litter into humus. The normal life of the earth's surface, in other words, can be maintained better by selective cutting than by clearcutting.

Bottom-line loggers are not likely to opt for the selection method simply to protect the soil from environmental extremes or to preserve natural balancing mechanisms. Selective logging, they aver, is too expensive. Each tree to be cut has to be singled out in advance. A road system has to be planned that will reach to every piece of timber, yet the roads cannot interfere with the trees that will be left intact. Tree fallers have to take special care not to damage the residual trees during their work. This requires skilled labor and extra time in preparation. Often, trees have to be rigged with cables or felled with hydraulic jacks to change the direction of fall away from their natural lean. Fallers, buckers, and choker setters have to cover more ground in their work, as do the cats that remove the logs. Since the machinery is decentralized over a larger area, there is a decrease in operating efficiency. Even the cleanup process can be longer and more costly; broadcast burning is harder to control, so the slash has to be bunched in small piles if it is to be set on fire.

These arguments sound plausible, but other variables tend to make selective cutting cheaper, in effect, than clearcutting. In a clearcut everything must be removed, including brush and undersized trees. This involves extra time, and time means money. Marginal trees are hauled off to the mill, even though there's little or no profit in them. In a selective cut, on the other hand, the noncommercial vegetation requires a minimum of handling, and only the valuable trees are taken away. Efficiency increases when all effort is focused on more valuable commodities.

A study of logging costs for second-growth ponderosa pine in California concluded that group selection was actually slightly cheaper than clearcutting, and that single tree selection was only slightly more expensive. The total variation in cost between the cheapest and the most expensive method was only 10%.[2] A federal study prepared by the Forest Service and presented to Congress concluded that the selection method was generally more expensive than clearcutting, but the difference was only a dollar or two per thousand board feet (the precise difference varied slightly, depending on the tree species).[3] When the price of stumpage is several hundred dollars per thousand board feet, a difference of a dollar or two in logging costs is not really significant. Management decisions must be made according to other criteria.

Economically, both clearcutting and selective cutting can be made viable in the appropriate circumstances. Most companies prefer clearcutting, but there are also those that have operated profitably under selective systems. Boise Cascade used computer programming on its 196,000 acres in southern Idaho, and the computer decided that the selection method resulted in optimal productivity.[4] In western

Oregon, the Woodland Management Company has made an impressive profit on cutover land by applying its own version of a selective system. In Santa Cruz, California, the Big Creek Lumber Company has out-lasted local competitors by selective cutting in the redwoods for almost half a century.

Bud McCrary

Lumberman, Big Creek Lumber Company

"Our family started lumbering commercially in 1946. When my brother, my dad, and I got out of the service, we formed a partner-ship with my uncle. We didn't have much land; we were buying timber from other people. We built a sawmill and moved around to different places.

"In the early days, we found we were cutting about 65 to 70% of the merchantable volume. Most people wanted to take off 70% of the volume and leave 30% or less, so it would come off the tax rolls.[5] But we never did get into any clearcutting or any heavy cut-ting. We started out with selective cutting, and we felt that was the best way to log, so we stayed in that mode.

"We also used smaller equipment than most people. We started out without much capital investment. Using smaller equipment, we found we were doing a lot less soil movement; so we stayed with that through the years. We found out that we could log cheaper with smaller equipment in some cases. Not cheaper in manpower, but cheaper from the standpoint of investment in equipment and fuel. Most people don't look at it that way. In the end, we spend more on labor, but we save money in equipment and overall costs. Today, it does cost us more money to log than people up north pay out, but we achieve better results by far.

"The way we practice good forestry on these steep hills is with labor-intensive forestry. We do things the hard way. We're not trying to modernize as much as we should, I guess, but we seem to be surviving. We feel it's important to give people jobs in this area. There's just a shortage of work for people, and if everybody tries to economize on labor, you run out of jobs for people and you've got a bunch of machines and nobody around to run them.

"You have to have darn good fallers to practice the kind of selec-tive harvesting we do, because you don't want to damage too many of the smaller trees. You can't just send some joker up there and say,

'Here's a chain saw. Go up there and fall that mountainside.' You've got to have somebody who really knows what he's doing. We also do a lot of cable rigging and jacking to pull the trees the way we want them to go. That all takes more time and more labor.

"We lay out our logging roads in such a way that the tractor doesn't always get very close to some of the logs. So we wind up with two or three choker setters behind the tractors. That gets expensive, but we feel it's well worth it. We feel that our careful tractor logging will actually create fewer erosion problems in many cases than cable logging. We think we've proven that, and people are beginning to agree.

"More than the cutting, it's the excavating that is really damaging our resources in the long run. We use small equipment and we use planning. We plan the operation carefully. I have three foresters working for me, and we only log about ten million feet a year. But I've got three foresters out there, both making timber contracts and managing the logging operations.

"In most cases the foresters with other companies lay out the original plans; then a logging superintendent comes in and logs it to get the lowest possible cost. In our case, the forester is responsible from the time the job gets started, through the execution of the job, to the follow-up for the job, which takes us another three years. Three years after we've finished cutting we're still in there maintaining erosion control.

"I've trained my foresters in erosion control. It took years of training, but they're able to do it now. I learned this the hard way, and I had to pass it on to them. You go in and size up an area and decide where you want to put your roads, first of all, and see whether or not you can get a road into the area. Then you start developing a road system that will minimize the amount of excavating you will do. That's the most critical part, and that's where you cut down your costs. If you can engineer your road well enough, you can reduce your excavating, which reduces the environmental impact and also reduces your costs. You minimize the width of the road on steep ground. If you have a D-6 and use it properly, you can do an excellent job.

"The real environmental damage, the damage to the soil, started when the D-8 cats came out around 1935. After World War II they started building stronger tractors, and we had more damage. Then out came the D-9, and we had even more damage. People just didn't recognize what was happening. I don't know why, because when we first started logging here on the Hoover ranch, we recognized immediately that the little gas-powered tractor we were using—equivalent to a D-4—made a lot less tracks in the forest than a D-7 or D-8 would make.

"Through the years we have stayed with narrow blades. And we sharpen our blades. We actually try to cut our roads with very, very sharp equipment. It's like a surgeon using a scalpel. If you make sharp and clear incisions, then you make very light incisions, whereas if you use a blunt instrument, it takes a lot more power to operate it, to push it through. The equipment doesn't handle as well, it's off. You don't build a road just exactly where you want it, so you're moving more soil to get the same results. When you use a dull knife, it doesn't go where you want it to go. The same thing with a tractor. If you have sharp blades, real dagger corners at the end of the dozer blades, you can put that thing right where you want it. That first pass goes right exactly there. You don't have to dodge off, go above or below a rock, you go through the damn thing. You'd be amazed at the kind of roads we build.

"We also build our roads so that a logging truck just barely fits on them. We figure there's going to be one truck on there at a time. We put turnouts every so often. You put the turnout where you have a wide spot so it won't damage the ground. We use those cheap Mickey Mouse radios, CB radios, and the truck drivers communicate with one another when they're going in and out of these roads. The guy might have to wait two or three minutes for a truck to come out, but we don't have the big highball operations, where the trucks have to pass at forty miles an hour. If you try to build two lanes of road around a steep, 70% sidehill, you're going to excavate probably eight to ten times as much as you would if you were putting a single lane in, a minimum-width single lane, because the excavation tends to square as you increase the depth and height.

"Another part of our operation is a plan for maintenance. You try to decide where your water is going to go, the water that runs off the road. If you don't really know where that's going to go, then it's hard to build a system that will stay in there. So the forester learns through experience where to put those water bars, whether to dip the road in or tip the road out, so the water doesn't run off the side of the fill. It always runs into solid ground. You try to dump your water into a stump, or group of trees, or a rock, some place that will slow up the water velocity. Once a person learns this, he's got most of his maintenance work pretty well done.

"Erosion breaks will sometimes fail; you get a little landslide or something will fall off the bank. Then we come into the area of winter maintenance. After a major rainstorm, we always send crews out in our new logging operations to make sure the erosion breaks are all in and operating properly. We do that for three years, and by the end of that time everything is pretty well solidified. You could have a problem after that time, but if an erosion facility lasts for three years, a lot of vegetation will be established by that time.

"It takes a lot of time and thought to do these things, and it is costly, but that's part of the cost of doing business, and doing it right. Of course there's no such thing as preventing erosion. Even though you leave it natural, the hillside is still going to get damaged eventually. A large, overmature tree will become uprooted under its own weight, and when it lets go, it takes out a big chunk of sidehill. You can minimize it, you can keep the erosion down to the natural rate, but there's no way to prevent it completely. We have a letter from Marvin Dodge, a Ph.D. who worked for the Division of Forestry, stating that his teams found that our erosion rate on two out of three jobs in the Santa Cruz area is less than natural erosion rates, and one of those jobs was just at what they considered to be natural erosion. Those were jobs that we did ten years ago. It shows that if you log properly, you can minimize erosion.

"When we put in a road system, we don't want it to wash away. We try to make it permanent. We try to build a road system that will act as fire protection for the landowner. We find today that people are hiking through these hills and camping anyplace they want to, and we're constantly getting fires started by campers out in the woods. I can't tell you how many times our logging roads have actually prevented a major forest fire. At least four times in Waddell Creek, forest fires have been stopped by immediate access by ourselves or the Division of Forestry. Once you've changed a forest to a second-growth stand of timber, you worry about fire, because you've got a succession of trees coming along and you don't want to lose that chain of succession. If you burn out all those six- to twelve-inch trees, you don't have that next generation coming along.

"Those smaller trees are the key to the way we practice logging. Every time we go in for a harvest, we're selectively opening the canopy. We're letting in light for the next generation. We'll go from a thirty-secondth of an inch of annual growth to as much as three-quarters of an inch within two years. That's even with some of the limbs removed, being brushed off when we took the neighboring trees.

"We selectively harvest 60% of the trees over eighteen inches d.b.h., and about 50% of the trees over twelve inches in some cases, depending on whether those twelve- to eighteen-inch trees need thinning. If they don't need thinning, we don't go down into the twelve-inch too much, but we do take them where we feel we need to establish some spacing. Actually, the twelve-inch and some of the smaller trees are the next generation. If you wait a few years, the next time you come around the twelve-inchers will probably be eighteen-inchers. We like to think twenty years is the best rotation. If you could get a sixty-year cycle going, in which you entered three times during that period, you'd have a pretty good thing going.

"There's a 'Head Start' program that's already occurred here, where the old-time loggers came in and clearcut and burned and got us started with second-growth redwood. That's a wonderful heritage, a good start on modern forestry. If you treat that properly, you have a perpetual yield going. We think that if you come in and take out 60% of the trees over eighteen inches now and come back every twenty years, you'll have a continuous supply of eighteen-inch trees. We've got that down pretty good. We establish good spacing in the forest and we have a nice-looking stand when we're done. We can fly over a cutover area two or three years after it's done, and it's hard to tell it's been cut.

"Many foresters from northern California say that what we're practicing down here is not good forestry, that it's really aesthetic forestry, not forestry for maximum growth rate. Maybe I'd agree with that, I don't know. We're not sure where our particular brand of forestry is leading us, or how much we are reducing the productive capability of the land by operating the way we do. We'll learn that over a period of time. We have a beautiful forest reestablished where the old-timers clearcut and burned, and now we're trying something entirely new: selective harvesting. We're getting good growth on the residual trees that we have, but I think maybe at some point in time we should think about clearing some of these areas and starting fresh. Maybe we should clearcut every couple hundred years and selectively harvest in between. It's a possibility, anyway. We just don't know.

———— " ————

Uncommon Forestry

Selective logging received a boost by the invention of lightweight hydraulic jacks, which can take trees that lean 20 or 30 feet to one side and fall them in the opposite direction. In 1975 Ray Silvey, a lifetime logger who once set a world record in ax-throwing by hitting 25 bull's-eyes in a row, invented a jack that weighed only 39 pounds but could tilt over a tree weighing 52.5 tons. Using Silvey's "Little Feller," an operator can apply 10,000 pounds per square inch of pressure with only one hand.[6] Any modern tree faller can now be a Paul Bunyan, pushing trees over backwards and landing them precisely where he wants.

The implications for selective logging are profound. Most trees lean slightly downhill; in the old days this meant they would have to be felled downhill. Now, trees can be felled away from streambeds and directed into areas where they will do the least damage to residual timber. On steep slopes they can be felled uphill, reducing damage to the trunks caused by ground impact. A sizable percentage of timber has traditionally been lost when large trees shatter on rough and hilly terrain, but uphill directional falling can practically eliminate the damage. In a typical logging show, this means increasing the usable timber by about 10 to 15%, or several thousand board feet per acre.[7] Logging operators receive higher profits, consumers can buy more lumber, and environmental interests are pleased when the damage to streambeds and existing vegetation is minimized by falling the trees in optimal positions.

Another boon to selective harvesting was the "Uncommon Forest Management Program" developed by Richard Smith of Portland, Oregon.[8] Under "common" management, suppressed trees are believed to be genetically inferior and are periodically removed to make room for their larger, more vigorous neighbors; under "uncommon" management, the suppressed trees are believed to be capable of further growth and are left intact, while the dominant trees are removed one at a time. The uncommon system is a type of selective forestry, but there is one important catch: the trees are selected not because they have reached an arbitrary, preconceived size, but only because they are stifling their neighbors. A "dominant" tree might be twelve inches thick or thirty inches thick; the important question is whether it is interfering with the growth of the trees around it. If it suppresses other trees, it will be removed.

The uncommon system is based on the ideas of a nineteenth-century Danish forester, C. D. F. Reventlow. In his book *A Treatise on Forestry*, Reventlow argued that loggers should harvest "all those trees which by unfortunate placement prevent their neighboring trees from making rapid growth."[9] Using this criterion on 427 acres of second-growth timber, Smith tripled the volume of wood in twenty years, while simultaneously logging off 3 million board feet of lumber. The growth rate on the tract averaged 872 board feet per acre per year[10]—and that compares favorably with what the large companies can get on similar sites that are clearcut, scarified, artificially planted, sprayed with herbicides, fertilized, and thinned. But the uncommon system requires a personalized approach to the land, for each separate tree must be periodically evaluated by on-the-spot inspection of its relationship with its neighbors.

Reventlow's philosophy is not without its critics. Some believers in selective forestry challenge his policy of removing only the dominant trees. This, they say, constitutes high-grading. Shouldn't we instead be taking the worst and leaving the best in order to improve the genetic stock?

Orville Camp

Natural Selection Ecoforester

"The word 'selection' is meaningless; without parameters, it's totally ambiguous. What criteria are you using to select? I spent some time on 160 acres up by Cottage Grove where they were doing 'selective' cutting, and I was astounded at what was happening. What they were doing was harvesting what we would classify as the stronger dominants and leaving the weaker members. Their argument was that these other trees that they were leaving would come back and grow quite well—which they will, but their genetic base is being lost there. I figure that if you remove your best trees you're losing about 25% of your productivity in a single rotation. I base that on an agricultural concept: if you had pigs, for example, and you bred only your worst ones, that's about what you would get. So I thought that was a catastrophe in terms of genetic base alone, and that's what most selection programs do.

"Nature has the only time-tested and proven selection system for the removal of trees. The key to sustainability is nature's own selection process. The natural selection that I use is based on Darwin's theory. When you have so many trees growing in an area and there's not enough room, the weaker members become selected out. What I do is try to learn the indicators for determining which trees nature is selecting out. Learning to read those indicators is actually quite simple; it doesn't take more than a couple of days to teach it. You start by looking at the top of the canopy, at the dominant trees. I have a golden rule that says: 'Never harvest the stronger dominants; only take the weaker members.' In a natural forest, the dominants will have the best genetic traits that have best been able to withstand the environmental extremes in every category. Those below them are being selected out.

"That's the simplified version, but you also have to understand successional order. If you have a madrone forest, for example, that has Douglas-fir coming up from below, you have to recognize that those Doug-firs belong there, that they aren't being selected out. That's why a simple understory removal is not the same as natural selection.

"You also have to recognize that each tree is contributing to the environment for the others. Just because a tree has been selected out, that doesn't mean you remove it; it only means that it becomes a candidate for removal. We don't actually know that we can take *anything* out and sustain the ecosystem over a long period of time. But in practice over the years, I haven't seen anything to suggest that we can't take out *some* of the trees that are being selected out.

"But you have to be careful with how much you remove, because you must maintain forest microclimate conditions—the water and sunlight for each particular site. You also have to maintain a habitat for every living thing. To determine the health of the forest, you have to look at the population stability for all the species. We pay particular attention to the secondary consumers, because they're the checks on the whole system. Green plants are the producers; they start the food chain that sustains all life. The primary consumers that feed on them are the bark beetles, for example. The secondary consumers are the woodpeckers that eat the bark beetles. If you lose the habitat for the secondary consumers, the bark beetles eat up all the trees.

"So you can't take out everything that's dead or defective. You have to leave some of those trees in order to pay your workers— your wildlife workers, your secondary consumers. You have to pay them if you want to get any work out of them. You pay them in the form of dead, down, and dying material. If you don't pay them, they won't do their work and, sooner or later, you lose the forest and *you* don't get paid. But if you pay them too much, there won't be anything to support your own economic activity. So it's a middle-of-the-road approach. How many wildlife workers do you hire to do your work for you? How many trees are you going to remove to make money for yourself?

"We generally take about 500 board feet per year of Doug-fir. But we're not just managing for Doug-fir. We also have hardwoods and a full shrub layer of huckleberry brush, which produces valuable floral arrangements as well as berries. And mushrooms. I understand that here in the Illinois Valley we produced something

Natural forestry: a wildlife worker

like two million dollars of mushrooms last year, and that comes off dead, down, and decaying material. And sometimes we take madrone burls, which can sell for up to ten or twenty thousand dollars apiece.

"All these other values disappear if all you have is a twenty-year-old tree plantation. There is no economics in conventional forestry; that's the reason everyone is going broke. I don't know of any conventional forester that's been able to sustain the forests—or themselves—over the long term. In this program, there are a lot of people who are sustaining themselves. And their net worth is increasing every year. From an accountant's viewpoint, if you put money in the bank, your interest will be related to your principal. If your principal is steadily increasing, you'll be able to make more money.

"Man has never been able to reforest economically. Man has been living off nature's forest. If you plant trees and you amortize that out for a hundred, two hundred years, there's no way you can economically do that. It'd cost you millions of dollars for each tree.[11] (And this is assuming you get a forest back, which often you don't.)

"Another reason why conventional forestry doesn't work is that there's down time. If you have an eighty-year rotation, about twenty or thirty years of that is unproductive. The real productivity of a forest is related to green foliage, and not just Douglas-fir trees. In order to sustain Douglas-fir, all those other plants and animals are necessary. But in conventional forestry, at the beginning of each rotation they purposely try to remove the green foliage, which means that the land is unproductive.

"Conventional forestry is based on anthropocentric philosophy: man is the center of the universe. Forestry is simply agriculture; there's really no difference. In Oregon, if you grow a Christmas tree it's called agriculture; if you grow it for saw-logs, it's called forestry.

"Natural selection ecoforestry is based on a different philosophy: it looks at the whole system. It's not agriculture by any stretch of the imagination. Agriculture is the science of taking plants or animals out of their natural habitat, providing for their needs and defending them from their enemies. Natural selection ecoforestry is based on not doing that, on leaving the natural ecosystem intact so that nature continues to take care of the living organisms, providing for their needs and defending from their enemies. There's no reforestation costs, no management costs in this program—we're basically paying nature to do all that.

"Natural selection ecoforestry has received a great deal of public acceptance. Virtually everyone wants to practice it. I don't find any private owners who don't want to use this kind of a management concept. Since 1981, we've set up somewhere around 25 or 30 places. Everybody wants to maintain a real forest. But the government claims it can't be done on a large scale, and the timber industry doesn't want it because they'd rather just take the rest of the old growth. The irony is, now they're going out of business, whereas we would have kept them in business.

"I do not believe that there's any other way of taking from the forest. If we're going to sustain an ecosystem, I don't see any other selection process that can do that. This isn't something that I designed, it's just something that I have begun to recognize: Nature has the only time-tested, proven way, and we have to have some faith in the way nature functions. We can't do any better than that. Our society thinks that we need to have control and dominion over every living thing. Well, we don't have that kind of control. We can't even manage ourselves, let alone anything else.

———— " ————

A Shady Operation: The Shelterwood System

The alternatives to clearcuts are many and varied. The notion of a "selective" silvicultural system opens up many questions: What percentage of the trees should be removed? And which ones? Should we take only trees over 18 inches? Over 12 inches? Should we take the dominants so that the suppressed trees might prosper? Or should we instead take only the weaker individuals, saving the best for their reproductive capacities?

In recent years, perhaps the most popular alternative to clearcutting among professional foresters has been the shelterwood method. Shelterwood cutting has come into its own as a response to repeated failures in regenerating timber using the traditional techniques of cut, burn, plant, and spray. Clearcutting is poorly adapted for tree species that like the shade, but it was supposed to work well for the shade-intolerant species like Douglas-fir. In many cases it did, but in other instances even the sun-loving Douglas-fir seedlings mysteriously withered away on sites that had been clearcut and burned.

Poor regeneration has been a particularly severe problem in the southern portion of the Douglas-fir region and on slopes with southern and western exposures. On a problematic clearcut site near Ashland,

Oregon, some Douglas-fir seedlings were planted in the direct sunshine, while others were planted in the shade of low brush, rocks, or dead logs. At the end of two years, only 10% of the seedlings planted in the open were still alive, while over 50% of those in shade had survived.[12] Total exposure was too great; moisture and temperature extremes had taken their toll. Similar results on other clearcut sites have forced some commercial foresters to change their minds; Douglas-fir may not tolerate shade, but it doesn't seem to like too much sun, either.

The shelterwood system is designed to produce partial but not total shade, to afford limited protection for sensitive seedlings. In shelterwood cutting the mature trees are removed in stages, and some semblance of a forest canopy is maintained until regeneration is well established. Typically, the forest is first prepared by thinning out defective trees and some of the noncommercial brush. Loggers can then remove a major portion of the mature timber—sometimes more than half, sometimes slightly less. Ideally, this initial harvest takes place just as the cones are opening during a good year for seed production. The seeds are scattered about and deposited in the ground as the logs and brush are dragged around the woods. After the loggers have left, the regenerative process occurs under ideal conditions: the forest floor is partially opened, but also partially protected. There is enough sun and moisture for the seedlings to grow vigorously, but not so much as to produce detrimental extremes in the microclimate. Fifteen months after seedlings were planted in adjacent clearcut and shelterwood tracts in southwest Oregon, 94% of the young trees in the shelterwood cut were still alive, while only 17% of those in the clearcut had survived.[13]

Five, ten, or perhaps fifteen years later, the seedlings begin to offer protection for each other; they act as a windbreak, produce their own shade, and transpire excess moisture from the earth. At this point, the assistance of the protective overstory is no longer required, and the "shelterwood" trees can finally be harvested. The forest then consists entirely of young trees, just as it does after a clearcut. The shelterwood system is therefore a type of even-age management, even though there is a short period during which there are trees of two distinct age classes. Shelterwood cutting resembles selective cutting in its maintenance of a continuous forest canopy, but it resembles clearcutting in its simplification of the ecosystem into an even-age tree farm.

There are several distinct advantages to the shelterwood system. When done properly, shelterwood gives natural regeneration an optimal chance of success. The parents of the new crop will be hardy, merchantable trees that have already passed the test of time by adapting to local conditions; indeed, the "leave" trees can be selected specifically for their desirable genetic traits. The seeds will find a partially disturbed soil in which to imbed themselves, yet this soil will not be subjected to the ravages of overexposure. As seedlings grow, they will be treated to just the right quantity of sunlight. Tolerant species can be given more protection than intolerant species, for the forest manager can adjust the amount of shade simply by changing the intensity of the cut. Since there is always some shade during the transition from one generation to the next, the seedlings are less likely to be overrun by pioneer, sun-loving species of brush. And, since there is less competition from brush, there is less need for the application of herbicides to counteract that competition.

Yet shelterwood cutting, like the other silvicultural systems, has its vices as well as its virtues. Because it is an even-age system generally applied to a single tree species, the shelterwood system lacks the natural stability based on diversity. Through most of the rotation, the stand structure is not truly mixed. Although a shelterwood system could theoretically include several crop species, in practice this is rarely done. As with clearcutting, it is more likely to require artificial props to keep the system in order.

As with selective methods, on the other hand, shelterwood harvesting is accompanied by the problems that stem from repeated logging over the same areas. It necessitates the building of a vast network of logging roads, and in rugged terrain these roads threaten the stability of the earth. The frequent operation of equipment compacts the soil, increasing the damage from runoff during storms and slowing down the growth of trees. Residual trees left after the initial cut are subject to windthrow and sunscald. Trees that are damaged by equipment or by falling timber become more prone to disease. And when the last of the sheltering trees are removed to open up the forest for the upcoming generation, many of the seedlings that have been so carefully nourished and protected are inadvertently but inevitably destroyed.

Another set of problems stems from the interaction between silviculture and politics. The success of regeneration under the shelterwood system depends in large part upon timing. The quantity of seeds produced by many tree species, including Douglas-fir, varies significantly from year to year. If the initial cut is made during a sparse year, the chances of adequate regeneration are greatly reduced. But the timing of the cut is often determined by political, not silvicultural, criteria; harvest

Shelterwood cut: a Douglas-fir overstory plays nursemaid to young western hemlock

begins when the plan is approved, not when the seeds are abundant. Regeneration failures would be a lot less frequent if the decision of when to cut could be left exclusively to on-the-ground foresters.

Politically, the shelterwood system can also be abused by those who want to bypass the restrictions on clearcutting. In California, shelterwood cutting can extend to a wider area than clearcutting if the forester declares that it will create "significantly less" disturbance. And if natural regeneration is inadequate after only two years, the seed trees can be removed—turning a shelterwood cut into a clearcut. Yet natural regeneration is more likely to fail precisely because the timing of the cut is affected by the regulatory process. *Shelterwood,* in this context, is more than simply a silvicultural term; it is an administrative loophole.

Despite its drawbacks and opportunities for abuse, the shelterwood system has some interesting possibilities. Imagine a program for a second-growth stand in which foresters had free reign to implement a shelterwood system based solely upon silvicultural criteria. They could time the initial cut to coincide with a good seed year for Douglas-fir, but they could also leave standing a few representatives of other species which occur naturally on the site. Then, instead of making the next cut in ten or fifteen years, they could actually leave the sheltering trees standing for a full rotation. When they come back sixty or eighty years later, they would be able to harvest genuine old-growth timber along with a percentage of the younger trees—but they would still leave enough trees standing to protect the subsequent generation. In this kind of leap-frogging scheme, a mixture of ages and species could be maintained at all times. Since some trees would be allowed to grow through two rotations, valuable old-growth timber would still be produced. Repeat entries, however, would be far fewer than with a selective system, while regeneration of shade-intolerant species would be easier. According to some scientists, this sort of modified shelterwood program could be the wave of the future in the Pacific Northwest.[14]

The Balance Sheet: Weighing the Alternatives

If each silvicultural system has both virtues and vices, how do forest managers decide which one to use? Sometimes the decision is made according to preconception or prejudice. A Douglas-fir forest in southwest Oregon, for instance, might be clearcut because all the textbooks say Douglas-fir requires sunlight for regeneration, and because it is company policy to clearcut for reasons of operating efficiency. Perhaps a steep slope in northern California, owned by an investor in San Francisco, will be selectively cut because the owner has been told by environmentally conscious friends that clearcutting is the eighth deadly sin. Basing decisions on such preconceived notions can have disastrous consequences. In the first case, the Douglas-fir seedlings may perish from overexposure. In the second case, the slopes may crumble under the burden of a massive road network constructed to remove a handful of trees from the middle of the forest. In both cases, decisions should be based on more specific information and more profound criteria. The specific sites should be evaluated first; only later should choices be made. Of course, there is always some evaluation of a site before it is logged, but often the evaluation is far from objective; it is made simply to satisfy legal requirements for decisions already finalized in faraway offices.

How can on-site evaluations determine the choice between clearcutting, selective cutting, or shelterwood cutting? Consider the following examples:

(1) In a remote area in the foothills of the Cascades, an epidemic of dwarf mistletoe, a parasitic plant, has broken out. Although the mistletoe is spreading rapidly, it is still confined to the northeastern slope of a ridge. The only "cure" for dwarf mistletoe is to remove the parasite from the treetops, and this is done most easily by harvesting the afflicted trees. Selective harvest would open up the forest canopy and damage some of the residual trees, creating ideal conditions for further spread of the mistletoe. To control the parasite and salvage the damaged timber, the area should be clearcut. Since the cutting will be done on a northeastern slope, regeneration is not likely to be impaired by solar overexposure.

(2) In northwestern California, forty acres of second-growth redwoods are to be logged so a family of urban emigrés will have enough money to build a house on their land. The area is populated with other small landowners, who have become accustomed to noncommercial uses of the forest, such as hiking and fishing. A primitive network of old wagon trails already exists in the forest, dating back to the time when the trees were first harvested around the turn of the century. What sort of silvicultural system should be applied? Selective cutting will preserve the noncommercial uses of the land. It can finance the construction of a house and provide for the possibility of additional income ten or twenty years hence. Construction damage will be minimal, since the old wagon roads can be opened up to service most locations. Regeneration will come easily, despite the existence of a continued forest canopy, for redwood shoots will sprout quickly from the stumps of the harvested trees.

(3) On a southern exposure in southwestern Oregon, a timber company wishes to liquidate its mature Douglas-fir timber and start a new crop of trees. On neighboring sites that have been clearcut, regeneration has suffered from overexposure, and most of the seedlings have perished. In order for these sites to be adequately stocked with Douglas-fir, they have been replanted several times and sprayed repeatedly with herbicides to control the brush. On other neighboring sites, however, single-tree selection has failed to produce an adequate stocking of young fir. There, shade-tolerant species of lower commercial value have over-topped the Douglas-fir seedlings. Since regeneration has failed from both too much and too little sunshine, some balance between sunlight and shade is clearly in order. A shelterwood system will provide an even balance of sun and shade throughout the forest floor, offering an optimal environment for seedling survival. If the land is not too rugged and can sustain the construction of roads and the repeated entry of the loggers, a shelterwood harvest can

simultaneously remove the mature timber and help establish well-adapted seedlings, while subjecting the earth itself to only minimal abuse.

Most management decisions, of course, are not so "clear-cut." Often, some aspects of a given site suggest one silvicultural system, while other aspects of the same site favor an alternate system. The variables must then be weighed against each other, trading off negative impacts with positive results. And the choice of silvicultural system is only the first among many decisions that will determine the fate of the forest. Will a road be placed here or there? Will the logging slash be piled, burned, or shredded? Will the brush be treated with herbicides, hand-cleared, or left as it is? Will the land be treated with chemical or biological fertilizers, or with no fertilizers at all?

Holistic forestry does have its preferences. It favors the harvest technology that least damages the soil. It tends to favor the silvicultural system that provides only minimal shock to the forest ecosystem. The holistic forester, however, has no pat answers, no absolute solutions. Each site must be evaluated separately; each problem is unique. The ground rules are simple, elemental. The health of the land is always paramount. The needs of the future must be weighed against the needs of the present. And, environmental engineering should be held to a minimum. When given the choice between two management options to accomplish a desired goal, the one that requires the least tinkering with natural processes is preferable.

6

HARVEST TECHNOLOGY

Logging, the actual harvesting of timber, is the most dramatic and immediately destructive aspect of forestry. The removal of giant trees is bound to have an impact on a forest environment, but that impact can be mitigated or increased according to the type of tools used to harvest timber. There is more than one way to skin a cat, and there are many ways, from horses to helicopters, to harvest a tree. In order to select the right tool for the job, we need some familiarity with the various forms of logging technology.

Big Cats in the Woods

Ever since it made its debut in the second quarter of the twentieth century, the caterpillar tractor has dominated logging. Cats make the roads that crisscross the forests and offer ready access to the trees that will be removed. Cats prepare the beds upon which the larger trees will be laid to rest; they also help control the direction of falling timber. The same machines then hitch up the logs and snake them out of the woods to the truck landing. Cats even clean up after themselves, pushing around the slash, piling up the brush to be burned or churning it into small pieces that will be left to rot on the ground.

With power to spare, the logging tractor can alter the very face of the earth. In the past, natural barriers had to be avoided; today, they are simply removed. In the old days, the falling beds had to be prepared with piles of brush; the earth itself remained where it was. The old-timers clearcut and burned, but they lacked the wherewithal to completely change the contour of the ground. When they built a road, they had to rely on their own muscle and sweat, along with the power of beasts of burden. The number of roads was therefore limited, while now the woods are honeycombed with roads. With caterpillar logging, between a quarter and a half of the available space can be taken up by roads and skid trails. When the cats are also used for site preparation, more than three-quarters of the earth can be disturbed by the movements of the ubiquitous machines.[1]

A logging tractor disturbs the earth not only with its blade but also with its crawlers, the ribbed tracks that grip the ground and give the cat its phenomenal traction and stability. Working under the weight of several tons, the ribs dig into the earth and churn up the top few inches of dirt, while the mass of the machine compresses and compacts the soil that is not overturned. Hoping to minimize damage to the soil while simultaneously increasing the speed of these mobile logging machines, operators in recent years have begun to switch from ribbed tracks to rubber-tired wheels.

Wheeled skidders have been heralded as environmental saviors, promising to relieve the topsoil of the adverse impacts of crawlers. They have, in fact, disturbed less earth and compacted the soil less severely, since they are generally not as heavy as the crawler tractors. But wheeled skidders have their own set of drawbacks. For one thing, they are far less stable. Many an operator has had a limb crushed under the weight of an overturned skidder. Lacking the power and traction of crawler tractors, they are not as useful for road building. Lacking the mass of crawler tractors, they do not serve well as anchors to control the direction of falling timber. Ironically, they can actually do more damage to residual trees during logging operations, for the logs are harder to control when skidders turn sharp corners.

Wheeled skidders might be modifying logging practices, but they are hardly revolutionizing work in the woods the way caterpillar tractors did a half-century ago. The cats opened up hitherto inaccessible regions, paving the way for a brief, short-sighted, and unplanned plunder of the timber resources of the backwoods. The adverse impacts on the forests were profound, but the social impacts of cat logging have been equally profound, and these are rarely talked about or even noticed.

In the old days of steam donkey logging, the workers had to be organized in large crews. The life and livelihood of each man depended on his co-workers and on the company. The fate of a worker might have been determined by a donkey puncher a half-mile away. If the whistle punk signaled a second too early, a lumberjack might have found himself cut by a cable, crushed by a log, or suspended in thin air. No woodsman was boss for himself; even the foreman had to answer to the company, which was run by capitalists in faraway places.

But the cat has changed all that, granting instant power and liberation to the working logger. Today, the single-man logging show is a living reality. A catskinner can fall and yard his own trees with no help from anyone else. Throughout the backcountry, small parcels are logged by individuals or small groups of men who are not dependent on others for their own safety. Even on large parcels owned by corporations or the government, much of the work is contracted out to small logging crews and independent operators. Land ownership has become increasingly centralized, but the existence of the caterpillar tractor has kept the actual work in the woods from becoming equally centralized. The owner of a cat has power over the trees and power over himself. He can contract out to work for others, but once in the woods, his world is simplified into three basic elements: the logs, the logger, and the logging tractor.

Joe Miller

Catskinner

"I was a mechanic in the [San Francisco] Bay Area, fixing cars for other people, fixing other people's problems. I wanted to leave the city, so my wife and I came up here. Our immediate reaction was to find a piece of ground to homestead. We walked around and found one with water and a lot of wood. We didn't own the land; we were just squatting. We proceeded to build a house out of a redwood tree that had fallen down and shattered. I saw what I could do with that one log, and I saw how many logs were lying around. I was impressed.

"We were living near the commune, and they had just accepted the Lord. So every time we came out of the mountains to be with people, we heard talk about Jesus. It started making things fall into place. There was no pressure, no shoving anything down our throats; so we accepted the Lord.

"I never lost my connection to the wood that was all around me. I started making fence posts and grape stakes. I started making shake bolts and taking them into town. Everywhere there were these big old chunks of logs laying around, and that was money.

"I started working on a truck. It had a winch to drag the logs out of the brush and load them on the back of the truck. Then I'd drive them down to where they could be made into fence posts, shake bolts, or whatever.

"By that time we were living at the commune, and as it grew I had less and less time to work on my own ideas. I was working more toward the benefit of the community, working on people's cars and fixing washing machines. It got to the point where I had to decide whether I was going to live my life full-time for Jesus, administering His gospel, going out to the four corners of the world. But I had this intense desire to go get those redwood logs. It sounds corny, I know, but they held a special attraction for me. It wasn't like working on cars, it was like panning gold. There the logs were, and people were just ignoring them. But I knew what I could do with them.

"So after living at the commune for about two years, we left, borrowed some money from my mom, and bought a four-wheel-drive truck. It was called 'The Frankenstein.'

"My tools were a chain saw, a hammer, wedges, blocks, and a piece of cable. I'd go out and cut the log to a size I could pull, and get it out in the road and make fence posts or grape stakes or whatever I could make out of it.

"I started hustling landowners to work their wood. Then I got my dad to put up the money to buy a piece of land. I got a World War II half-track with a winch on it, which had more traction than the four-wheel drive. And I got another chain saw.

"And then I got my first cat. Sold the half-track, sold the truck. The cat was really just the thing. It didn't slide. You didn't get stuck. You could turn on a dime. You could pull anything you wanted—and it didn't break down. It was just a hunk of iron made to order for pulling logs out of the brush.

"It was a really old cat. If I didn't keep it loaded with Power Punch, it burned eight gallons of oil a day. But I kept it so glued together with Power Punch, which is like STP, that I got it down to two or three gallons a day.

"Every time I looked around there was more and more wood laying around, more wood than I could possibly work. So I discovered how to send logs to the mill instead of working them myself. I had to go to Eureka to pick up some machinery I had up there, and I rolled a few small logs on the truck and took them to the stud mill. I was very upset that they didn't pay me when I delivered them, because I was living hand-to-mouth. But when I got the paycheck, I was thoroughly shocked, because it was so high. It turned out I made more money with less work. So I started hustling as many logs as I could to the mill.

Power to spare: all-purpose logger

"About that time I started doing 'green' logging. I figured that if there was that much money in sending the logs to the mill, I might as well start selling green logs, so I could pay my property off. So I started falling trees, and I had to buy a loader to put the logs on the big log trucks.

"I didn't have anybody working with me. I'd be jumping in and out of the cat, which is some of the hardest work there is. On the hillsides it's so steep and it's so high up to the seat that you find yourself climbing around like a monkey. You park the cat, take the brake off the winch, and pull out this one-inch winch line. If the log is downhill, you can lean into the cable and turn the winch, but if the log is uphill—well, then you really have some work to do. You hook the line up to the choker, which is hooked up to the log, and then you go back up to the cat and put that winch in gear and bring it in. Then you build your turn of logs, hook them together, and take them down the road to the landing. All this time you're getting on and off the cat, climbing up to the seat, and jumping down again.

"So here I am green logging, doing it all by myself, and finding it's very slow. So I started thinking about employees; but employees meant ties with the government, insurance, and responsibilities. I didn't like that, but I did it anyway. I went out and got workmen's compensation and filed with the state and the federal government.

"The first guy that worked for me was a good worker, and I enjoyed him, but he wasn't always there. He lived off in the woods and didn't have a phone. A lot of times he just didn't show up for one reason or another, and there was no way to reach him. Since then I've had sixteen employees, four or five at a time. But it just hasn't worked out. Motivation is the biggest problem. Motivation to come to work, motivation to work, motivation to do good work. And motivation to enjoy one's existence.

"From now on, I intend to work mostly by myself. I want to work with my ideas through my own hands. I don't want to be pushing my ideas through someone else's hands, because that doesn't work. I can't seem to motivate anybody to do it the way I would do it. And I get more enjoyment out of doing it than getting someone else to do it.

"My machines are my best employees. Especially my cats. I can manipulate them. I can have them do whatever I imagine them to do. I can run that machine to its maximum. I know how it runs. I know how much power I can use. I know when it needs grease, what parts are moving freely and what parts are sticking.

"The machine becomes an extension of your hand. All the time you're seeking its limits. Of course machines break down if you take your mind off what you're doing. When you're sitting there grabbing the levers and cussing at it, it's going to break down. But if you're sitting there feeling and flowing with the machine, it doesn't break. It works for you, because it's just a machine. It's not out to get you.

"I love working on my machines. I have a shop and there's a proper tool for every purpose. If you don't have the proper tool, it's frustrating; but I know that, so I don't mess with it. I get the proper tool or figure out a proper substitute. There is a right way to do everything. If you sit there and think about it long enough, you'll figure out a solution. It's creative. You find out new ways of doing things, and when they work out real slick, they make you feel good at the end of the day. Sometimes the work takes a month or two, but then the finished product rolls out: slick paint job, brand new decals, and nobody else's Mickey Mouse fixit jobs in there. You've gone through and freed everything. You've cut loose all the things that were welded and you've fastened them properly with bolts or pins or whatever. All those sloppy joints, you've shimmed them up. And you feel shimmed up. You feel together.

"Caterpillar tractors are designed to run forever. They just don't wear out. You can take out the bolts and put them back in every ten years for a thousand years or whatever. Take my 1946 cat. Some guy must have known just exactly how much steel to put in everything. It's just a good tool. In my business, it's the best tool.

"I use my cats for everything. I even fall timber with a cat. If I need to knock out a high spot to make a bed, the cat will do it. If I need to pull the tree, the cat will do that, too. If you run into any problems, you've got a cat there as one of your tools. With a cat you can line all the trees up in a straight line, so you don't have to sweep the ground with your logs. You can fall the tree any way you want to. You cut your tree and leave your hinge, but you make that hinge extra thick. You set your wedges in the back cut, then climb in the cat and pull it over center and there it goes. You get back out, buck it, limb it, and choke it. Then you drag it out with the cat. The cat does it all: it builds the roads, falls the trees, and hauls out the logs. It's the greatest tool I can imagine.

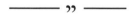

One man's virtue is another man's vice, and the logging tractor has led to rags as well as riches. The cat's power and versatility have enabled it to do the job of many men—and simultaneously put workers out of their jobs. Joe Miller has found he likes to do the whole show himself, but not everyone has $150,000 to invest in a new logging tractor or the ability to make an old one run like new. For lumberjacks who are not their own bosses, the caterpillar has cut a road directly to the unemployment office.

If used correctly on gentle terrain, the caterpillar tractor can be an invaluable tool. But the cats have been overused. Their power is just too great—and their abuse all too common. Hills are pushed around by catskinners like sand in a two-year-old's sandbox. The wounds to the earth are not consciously inflicted, but when such enormous power is at your command, it is easy to get carried away.

Even when used properly, the damage due to compaction can be extensive. According to a study of several sites in southwestern Washington, areas that had been logged by tractor (not including the roads themselves) showed a mean loss of 35% in soil permeability.[2] When soil becomes less permeable and more dense, trees will not grow as well. In a ponderosa pine plantation in California, the bulk density of the soil increased by 30% on skid trails, leading to a 55% reduction in tree volume after 16 years. On landing areas the damage was even worse: a 43% increase in density led to a 69% reduction in volume. Even on areas adjacent to skid trails, an 18% increase in density led to a 13% decrease in growth.[3] In a Douglas-fir plantation, 25% of the area remained heavily compacted 32 years after logging—and the net loss of tree volume, calculated over the entire unit, was 11.8%.[4]

But what if the cats did not have their destructive powers? What if these mobile yarding machines could simply pick up their tracks and step over the small irregularities in the earth's surface, instead of driving relentlessly over every little obstacle and packing down the ground? Indeed, what if they made logging tractors out of flesh and blood instead of steel?

Joan and Joaquin Bamford

Horse Loggers

Joaquin: "I started out as a choker setter for Weyerhaeuser back when I was in high school—1959, I guess. Then I worked up to rigging slinger, and then I was a hooktender. Then I ran cat and ran a skidder. Then I worked on a yarder. And I worked as a foreman. I've done many things since then, but primarily everything has been in the logging business.

"Riding back from work a few years ago, this guy says, 'You know, we ought to get out of this business.' I said, 'Yeah, I suppose we should.' I wasn't happy with the outfit I was working for. He said, 'You ever thought about horse logging?' I said, 'No, not really.' He said, 'Well, we should look into that.'

"I've always believed that if you want to look into something, you go to whoever knows most about it. I heard about this fellow Al Merrill. He was about sixty-five at the time, and he was still out there logging. I went out to meet him and went around with him, helping him out when I could, watching him and asking him questions. Of course, the first thing he said to me was: 'If you want to go into horse logging, you might as well figure on starving to death.'

"So we set out to get our finances all squared away before we started. You don't want a lot of payments to make if you're in horse logging. You need to have a little bit of money in the bank. And you need to have an awful lot of experience behind you, not only in logging, but in horse psychology, horse sense. What to do when a situation arises. How to move. How to be light on your feet. You have to be able to perceive what's going to happen to you fifteen, twenty feet down the road. You have to be able to jump out of the way, stop them, turn them. Think what the horses are thinking: whether they are spooked or scared; whether they'll stand for you; whether they'll back up for you if you want them to. If somebody else is out there, make sure you're never in a bind in case they happen to say a word that means for the horses to move ahead or move back.

"A workhorse adapts easily to logging. If he's used to pulling things, it only takes an hour or two to get used to working in the brush. But adapting to the person is the important thing. Bill [one of their horses] wouldn't even go into the truck until he had confidence in Joan. But he sure has confidence in her now."

Joan: "Bill's never refused to pull for me. When he gets excited, he'll go uphill pulling these great big ones sometimes, and he'll be digging in the ground with all fours, and you can just see his muscles working. But you should never put your horse on something he can't pull. That really ruins a horse. Makes him balky.

"Every time I take a log to the landing, I always let Bill rest. Because I really appreciate what he's doing for me. He enjoys the work. I enjoy the work. It's not for the volume that we do it. It's a challenge. Every log you take down is a challenge. You have to figure out the angle, if it'll bang up another tree, if it'll get you as you go around the corner. There are so many things. Every time you get down to that landing, you feel you've really accomplished

Modern horse logging

something. Then you come up and you take the next one out. It's always exciting. It's never dull. But it's so quiet, so peaceful. And it's so nice to be working with Bill.

"I've tried driving a team, but they were too fast. . . ."

Joaquin: "At five-foot-two, you know, she couldn't see around two rumps. She was running between horses, trying to see over them or around them."

Joan: "I'd drive them into a tree. There would be a tree between the harness and everything. One horse, you can angle him to where you see where you're going. I don't think I'd want a team now. Not after pulling with Bill."

Joaquin: "But you can't log as big a log. I mean, with a team you can log some really big logs. Most all the horse loggers I know have two horses. Not that they always use the two horses, because when you're thinning small timber, it's better to use one horse. You can get between the trees better, and you don't have to make a real wide skid trail."

Joan: "Joaquin had a neat relationship with his team. A lot of horses won't stand in the brush, but he had them so well trained, they would stand there all day. If a tree got hung up, the horses would pull the tree down for him."

Joaquin: "If you get a team that's easy and slow and pays attention and is real good at it, then you're going to be a success at horse logging. If you get a team that wants to run, that's a little bit jittery, that jumps onto the turn or doesn't work together, then you're going to get yourself hurt. We've had several horses that we had to get rid of because they got nervous and excited and didn't pay attention to you.

"The thing about horses is you can't get in a hurry. No way. Whenever you start getting excited and worried about getting enough volume in, then something's going to happen. You can't hurry the horses. You just have to take one step at a time. It's not like machine logging. When you own machinery and you have big payments (I mean they don't give that machinery away), the logger has to rip and tear and go right after it. There is no being really careful. He's got to get in ten loads a day, or whatever it takes to make those payments.

"In horse logging you have to look at the timber before you can go and say, 'Yeah, I can log that.' The best type is second growth, timber that ranges from fifteen years to forty or fifty years. You want it small, but not too small. We usually won't go into timber that is less than twelve inches on the butt.

"And you need to look at the land. You want a downhill pull, if at all possible. You look at your terrain. You find out what the soil is like: whether it's loose soil, whether it's on a steep hillside, whether there will be erosion where you drag the timber down. You have to look at it as a conservation practice.

"When you start out on a job there's a lot of preplanning. You go in first and clear the brush out of the way. You have to take off the bottom limbs so the horse doesn't gouge himself or hurt himself. You cut all the vine maples and the underbrush down low enough so when the horse steps on them he's not going to hurt his feet."

Joan: "Some people don't do all that. They let the horses work in the brush. They don't prepare the way for them. They leave limbs sticking out. Those horse loggers often have big scars on the horses' legs. It's a lot of work to clear the way, but we're doing it for the horse. We don't want to hurt a fine animal like that."

Joaquin: "About three years ago, a lot of individuals went into horse logging without the proper equipment or the proper horse. They had chains for tugs; or the collars didn't fit quite right; or they didn't have a pad. A lot of things were just thrown together.

You know, living off the land and having a horse and doing horse logging—boy, that's right back to nature. The people would go in and log properties where they really didn't have any experience in falling timber or working a team. They ruined a lot of teams and made a lot of people in the area upset with horse loggers."

Joan: "Remember the guy we saw who had a large saddle horse that he used as a draft horse? He wasn't built to move timber like that, but they used him."

Joaquin: "His bone structure wasn't really heavy enough; his muscle wasn't heavy enough; he didn't have the chest on him to do that type of work. He was a big horse, but he just wasn't a draft horse."

Joan: "The first thing you have to know is how to relate to animals. You have to realize what your animal is thinking at all times. People think they're dumb, but they know more than you give them credit for. In fact a lot of horses are more intelligent than the people I've seen trying to run them. It's amazing what a horse will do if you just get the communications squared away."

Joaquin: "You can get a bond with a horse. Like for lunch, while we have our peanut butter and jelly sandwiches, he gets his oats and grain. They really appreciate that. They appreciate having a blanket put on them after they're hot. They appreciate the hay they get. They appreciate getting some praise after they get down to the landing with a big log. They just eat that up."

Joan: "And if you're upset, you can transmit it to them."

Joaquin: "If you go out there and you're afraid, then that transmits to them. If you're not sure, then that makes them spooky. Because they know that you don't know. They can tell by your voice."

Joan: "Animals have their own personalities, just like people. Bill, he wants to go to work. He gets excited when he gets his harness up."

Joaquin: "When you get his harness, you can almost see him grinning."

Joan: "Because he knows he's going to be used. And then when you bring him back, he's really excited by that, too. You put him out in the pasture and he rolls and he just relaxes. He's glad to get home, but he's glad he worked."

Joaquin: "Again, you have to relate it to people. People want to be used. They want to do something. That's the way Bill is. He was bred to work. He doesn't want to be worthless.

"I'm the same way. I like to work. I love to fall timber. I make a
real game out of it. I really enjoy pulling a tree and driving in a stake
and trying to get it to go exactly where I want it to go. It's a real
challenge. It's exciting to be able to fall timber where you want it—
most of the time. If anybody tells you they never made a mistake
falling timber, they're either lying to you or they haven't been in
the brush very long.

"Falling timber is something you can do anytime. You don't have
to wait for good weather. Like today it's raining, so this would be a
good time to cut. Between falling timber and horse logging, you
can keep working year-round. This time of year, horse logging will
do a cleaner job than a cat. When the ground is wet and the cat gets
into a pull, the cat tracks spin and immediately start to tear the soil
up. Water will run down that and cause erosion. When you lock a
track, it spins and compacts the soil to make a rut. With a horse you
don't have that problem. Of course, a good catskinner working on
dry soil doesn't do things like that."

Joan: "But the cat runs over the little baby trees, the new growth;
and the cat can't get where a horse can go, between the trees. In
horse logging, you want to leave it as natural as possible."

Joaquin: "Ninety-five percent of everybody who's got timber, if
they're going to do any logging at all, would like to see at least a
couple of loads horse logged. So there's no problem getting jobs.
People want horse logging because it's interesting. They want to
see what it's like. They want to see a horse work.

———— ” ————

A horse doesn't weigh as much as a cat. It doesn't push around the
topsoil. It doesn't spin its tracks, create erosion channels, or significantly
compact the earth. It doesn't bump into residual trees. All in all, a horse
is gentler on the forest than a logging tractor or a wheeled skidder. And
a horse consumes renewable resources—hay and grain—rather than
unrenewable fossil fuels. A horse is not made out of metals that must be
extracted from the earth. It is not manufactured in factories that dump
polluting by-products into the air and water. Environmentally speaking,
a horse is preferable to a cat in every respect. And it is also cheaper
to buy.

For thinning small timber or harvesting poles, a horse can hold its
own; but try as they might, even the mightiest draft horses can't tug on
a large log as well as a cat. Horse logging is slow and tedious work.

When the timber is small and scattered throughout the forest, environmental benefits can offset the relative inefficiency of horse logging. But when the timber is large and most or all of the trees are to be harvested, the horse will be competing out of its league. Horse logging, in short, is a valuable tool, but it is not a panacea. Like any good tool, draft horses should be used only on the jobs for which they are best suited.

High-Leads and Skylines: Cable Yarding

A horse might be out of place on a clearcut of mature timber, but other alternatives to cat logging are specifically geared to such concentrated operations. Cable systems of various sorts have been modernized since the days of the old steam donkeys. Modern, gas-driven yarding machines with portable steel towers have replaced the donkey engines and the spar trees that once supported the cables. Today's portable yarders can be driven along logging roads, stopping at convenient locations to extend their steel cables thousands of yards into the woods. In the *high-lead* systems, the cable is hooked on to a log, suspending the front end while the tail end of the log drags and bounces along the ground. In *skyline* logging, a cable is suspended not only from the yarding tower, but also from a high point at the tail end of the system. A carriage runs along this elevated track, and a log attached to the carriage can be reeled in to the landing without ever touching the ground.

Cable logging has enjoyed a renaissance in recent years, largely because it is supposed to be gentler on the earth than cat logging. Since the cables commonly reach a quarter-mile or a half-mile into the woods, the need for logging roads is greatly reduced. And since the machinery is confined to the roads, there is less soil disturbance and compaction. There are numerous studies that compare cat skidding with cable yarding, and they all point to the same obvious result: a logging tractor zigzagging through the woods will do more damage to the earth than a cable dragging a log (high-lead), and the least damage is done when both the cable and the log are elevated (skyline). One such study in the Oregon Cascades concluded that tractor logging created three times more compacted soil than high-lead logging.[5] Another study revealed that a logging tractor severely disturbed 36.2% of the site while leaving 26.2% totally undisturbed; a comparable skyline operation severely disturbed only 2.8% while leaving 74.8% totally undisturbed.[6]

Another virtue of cable systems is that they generally yard the logs uphill, whereas tractors function best when skidding the logs downhill. The landings for cable systems, therefore, tend to be situated on

A spar tree of steel: the mobile yarder

ridgetops, where they do little damage; tractor landings, however, are often near streambeds, where the displaced earth is easily washed into the major waterways. Logging trails that lead to a ridgetop landing actually tend to spread the flow of water in several directions, for the water is diverging when it follows the trails back downhill. By contrast, logging trails converging on a streamside landing tend to channel all runoff into one central area, intensifying the damage due to erosion.

Cable logging does have its drawbacks. The landing pads tend to be large, for they must accommodate not only the logs but the yarding machine as well. When mobile yarders use the roads as landings, the roadbeds have to be significantly wider. The extent of soil disturbance increases with the square of the width of the roadbed, so even a slight increase in the size of logging roads is likely to take its toll. A study of several logging sites in California found that cable yarding was preferable to tractor skidding on gentle or moderate hillsides; but on steep slopes with a 70% grade or higher, "cable yarding produced from two to twenty times as much erosion as tractor operations."[7] Ironically, cable yarding is often used on steep or unstable hillsides precisely because it requires fewer roads and is alleged to produce less erosion, but it is on these very hillsides that the larger landings and haul roads will cause the most damage.

Then there is the question of cost. High-lead yarding runs about four or five dollars more per thousand board feet than tractor skidding; a skyline system generally costs about twice as much as logging by tractor. The extra cost stems from two factors. First, the capital expense is higher: a high-quality, versatile yarder costs several times as much as a logging tractor. The operating costs are also higher, for a cable system requires a larger crew. The cables must repeatedly be set and reset in new locations; this takes time, labor, and, therefore, money. In a skyline system, the time and energy required to reset the lines can be particularly formidable.

Traditionally, cable yarding has been difficult to adapt to silvicultural systems other than clearcutting. When a portion of the overstory is to be retained in a selective or shelterwood harvest, it can prove troublesome to lay out a quarter-mile stretch of cable that will not run into residual timber. A logger who harvests only those trees within easy reach of the main cable is "strip-cutting"—he cuts long, parallel swaths through the forest, while leaving the timber between the swaths intact. Strip-cutting is, in some ways, a viable alternative to clearcutting, but it is not truly "selective" forestry: the choice of residual trees must be based on an

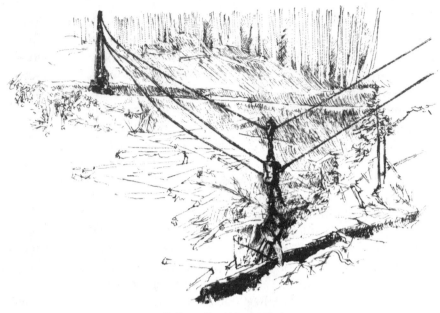

Skyline yarding with a grapple rig

arbitrary geometric pattern rather than biological or silvicultural criteria.

Alternately, a logger can keep his pathways narrow while reaching sideways into the woods with special tag lines attached to the main cable. In this manner, he can reach almost any tree in the forest without removing its neighbor. Until recently, this type of selective cable logging has required a sophisticated system with a locking carriage on the mainline, and it was consequently quite expensive. But in the past few years, a small crew of cable loggers from Ashland, Oregon, have practiced selective cable logging economically, using only a minimum of basic equipment.

Wally Budden
Terry and Vicki Neuenschwander

Cable Loggers

Wally: "I was in the parachute troops in World War II. When I got out of the army, I came over here to Klamath Falls and went into falling timber. Things got kind of tough, so we moved over to the coast. We didn't know nothing about falling redwoods, but we learned the hard way. I was young. We partied a lot and celebrated. We got into this salvage operation, and one month me and my partner made $30,000. That was a lot of money back in the fifties.

"We worked for Georgia-Pacific for quite awhile, and then we went out and bought some private timber. We found this one patch that was second-growth redwood, coming out of the stump. It wasn't big timber, but it was real thick. It was real steep ground. I had a young fellow there that had worked in high lead, and he knew all about how to rig. We went out and bought an old yarder. We raised the pole about a hundred and thirty feet, and that was quite an experience for me. I was learning as I watched. That's how I got into the cable end of it. When we got done it was wintertime, and I decided to build a yarder. I built one of the first mobile towers that was ever built.

"This one guy, Earl Shipley, he said, 'You know something, if you could learn how to get slack off a three-drum yarder, you'd make a million dollars.' Well, that kept going through my mind. And when I came up here, Terry [Wally's son-in-law] and I got to be partners. We went out and bought this old swing yarder that's setting up at the house now. We had a terrible time with it, high leading. We couldn't get slack and on this steep ground and everything. But we logged with it for three or four years. I knew what I needed but I couldn't find the answer, and all of a sudden a little choker setter said to me, 'Wally, why don't you take your haywire and go back through your carriage in reverse and molly it in on your mainline. Use your haywire for a slack puller. That way you won't have to drop the skyline and raise it.' This is real important information, because nobody on the Northwest is doing it but us.

"But anyway, we tried it and it worked. It's the slickest thing you ever saw. It's so simple; I think that's one of the reasons we didn't see it. We go back and hold the carriage in place. We can lateral yard a hundred and fifty feet; we can go out there two hundred feet if we want. We can bring it straight over, straight up, and straight in. It's your rig-up that counts. It don't take the power; it don't take this big equipment. Lift is the most important thing there is on cable logging. If you're dragging it on the ground, you hit a stump or something, you start tearing things up. You have to get lift in order to keep the damage down.

"Quite a few years ago, I had a choice. I had the expertise to become a big yarder logger. I could've bought a big slackline yarder and probably could've made a fortune with it, because they were putting up clearcuts. But I was upset by the devastation that was going on. I could see the tremendous road-building systems. I got a friend here, he's built six hundred miles of roads that's never grown a tree—roads that're going nowhere, but they cost the government millions of dollars.

"So I decided to get into the environmental end of this thing. I abandoned all of that big stuff. I said to myself, 'Someday, somewhere along the road, somebody is going to put a stop to it. They can't go on forever.' My wife kept asking me, 'When's this going to happen?' I said, 'Someday, it will happen, but timewise, I can't predict it.'

"In the meantime, we struggled, and it was hard on us. I decided to organize a small crew, and that's what Terry and I have been doing ever since. He's been a part of it."

Terry: "I do most of the climbing, the loading. Sometimes if Wally has to go down into the brush, I'll run the yarder, or I can help set chokers."

Wally: "Even now, you're only looking at a three-man crew. Terry does the hard work now. He's younger than I am and his legs are stronger. And he's good at it, too."

Vicki: "I haven't been working in the woods lately, but during our slim years I was a real important member because we couldn't hire anybody else. I did everything. Terry had to go get a trucking job for us to survive. My dad and I did four acres of ground for BLM. We had to make enough money to pay taxes that year, and him and I did it. He felled it all, I yarded. We cut firewood off of it, and Terry would come up on the weekend and just pick up the loads. That was the way I first learned how to run the yarder. Daddy said, 'If you can drive a car, you can run this yarder.' Well, it was never that simple, but I did it. I also set chokers, unhooked in the landing, whatever. Never driven the cat, but I have driven the loader. Basically, a woman can do all of that. We had five or six slim years where I worked in the woods."

Wally: "Logging has been a roller coaster all these years; it's up and it's down. It all hinges on the demand for wood products. You have your good times and your bad times. And if you don't prepare for the bad times when you have your good times, you won't survive your logging. That's part of this whole scenario."

Vicki: "I remember the timber crunch in '80 and '81. It started with slow housing. When the work came back, all that was left was big jobs, big equipment, big everything. Production. Everybody was so behind with timber, and the mills were so behind, they needed logs, and they needed them now. There was no supervision. It was a lot easier to give a ten-million-foot job to one logger than to give ten loggers one million. Then all they had to do was put the outside boundaries: 'Go in, take her all. We don't care what you do.'"

Terry: "Just as much paperwork for a small outfit as for a big one, so they might as well give it to a big one. Save themselves some work."

Vicki: "So when we went back and couldn't put out the production that they needed—were talking 'Let's save those nice little trees, those thirty- and forty-year-old trees'—they didn't care about that. They had just done their ten-year plan: clearcuts, and don't worry about anything.

"Finally, right at the end of the eighties, they began to look around and say, 'Oh, no.' The worst part of it was that they couldn't get the timber to come back. Right here in southern Oregon, they found out that all those clearcuts, they can't grow trees anymore. Now that they've found that out, ten years later, it's too late; they've lost that ground. They just spray it all."

Wally: "Just a few years ago, some people started to realize they had been doing it all wrong. That's when business picked up for us. We've got it down now to where we're pretty important. We're sitting in the driver's seat; we're in demand.

"The way we do it, it staggers your imagination. I bought this little yarder, it cost me a thousand dollars. We put another motor in it. We raise an A-frame; we hang all our rigging on it. Then you have to have a tractor, and either a triple-drum or a little sled yarder. Then the rest is all more or less stump rig. We have an awful lot of options that we use, but it's hard to explain. I would think, in order for anybody else to do this, they'd have to come and watch us, 'cause nobody else is doing it. Or we'd have to go and set 'em up an outfit. I'd be tickled to death to show 'em how to log this kind of way. We'd we happy to help 'em out as much as we could because we know that this is the way it should go.

"It's not too expensive in terms of equipment. Most of these little loggers have a cat, and they have a little front-end loader, or a little shovel. All they'd have to add to their equipment would be a little yarder of some kind, a triple-drum or a little sled yarder. That's all they need. We don't have a lot of money tied up in equipment. That's why our price is so reasonable. We're cable logging in steep, rocky, rough ground for about the same price they're cat logging. All we want to get out of this is a living.

"The quality of our work is hinged on one thing: keeping it small. We don't put out much volume, about two loads a day. But we're pretty consistent, we get 'em every day. It's making us a living, and we feel that we're doing something that should be done. I've talked to other loggers about it, and I've heard them say, 'You're setting logging back.' They're so gung-ho to buy these big clearcut machines and make a lot of money that they couldn't see doing it a simple way.

"Our road building is minimal. We don't put enough roads in to stick in your eye. Everything we do is hinged on the environment. If we had a lot of little loggers out there doing the same thing we're doing, it would put a lot of people back to work, it would do a nice job, we'd have a forest left when we're through. It'd all work perfect. That's the answer to the whole situation."

Vicki: "There's a lot of guys out there that know how to climb trees and do the rigging. The big deal with Daddy and Terry is that they're not using one of the new carriages. That's a big expense. The new carriages are actually motorized, and that's what pulls your mainline down. A little logger getting into it now would have to learn how to get slack out of just a regular, straight carriage, because if you're going into it small, you couldn't afford the new ones. They run up to around $350,000, maybe $500,000."

Wally: "And where does it get you? They do just the same thing we're doing with just two sections of haywire."

Vicki: "The problem is, most loggers aren't interested in this because we're limited in terms of production. There aren't that many small jobs out there. I've gone to the BLM bidding, and there'll be ten small loggers bidding for just one small job, maybe 750,000 board feet. And that was all there was for that bidding session."

Wally: "If they set these jobs up, I know in my own mind that the loggers will do it. They'll figure out a way. But they got to set up these small jobs for these kinds of people. That'll solve it right there, if they get these bureaucrats to quit looking at these huge, massive clearcuts.

"We don't just depend on the government; we work for the companies too. Right now, we're working for Weyerhaeuser; we're working around an eagle's nest. We're the only outfit in the country that they'll allow to go in around those nests. I had a big old eagle sittin' in a snag the other day, watching us with our little Ford motor. That little motor don't make much noise.

"We done Doak Mountain up there last winter, the largest concentration of eagles in the world. We done the whole mountainside. We took over two hundred loads off that. I have talked to person after person who said they didn't even know we were in there, after we got done. This might sound like an ego trip, but it isn't. We're the only people in the Northwest that can do this kind of logging. Downhill, we were logging probably about 1200 feet down off this rocky mountainside, because of this little deal: knowing how to get slack off a three-drum yarder.

"Awhile back, I worked for Boise-Cascade. I was environmentally logging for them, doing a beautiful job. They had this system: take a job and leave a job. Then you come back in thirty years, forty years—there'll be another job there. That's just about as simple as you can get. That's a sustained yield, and they were doing it. Then all of a sudden, Japanese needed logs, so they got greedy. They went back in the area that I had logged. I went back, and here was three big cats tearing up everything that I had worked so hard to make look good. I was so upset, I haven't worked for them since.

"Here's how I look at it: I love logging. This is my life. I'm not going to destroy the very thing that I love. When I die, I'm going to die out there on a stump somewheres."

Vicki: "I can remember as a little girl my dad saying, 'How many jobs is it that a person can go into the woods every day and have a picnic?'"

Wally: "If you spend your life out in the woods, you get to appreciate nature's way of doing things. When I'm falling timber and I get a little bit tired, I set down and sometimes I watch a bunch of little ants. There was a whole string of them; they were pulling these little fir seeds into this anthill. Well, you know how an anthill is; it's all fertilized. I was really curious about that, and I went back two years later and here was all these little trees growing out of there. I thought to myself, nature's got a system where they've got animals planting twenty-four hours a day. All the little squirrels, and the birds—nocturnal animals, too. How can man compete? People talk about how they're planting six trees for every old-growth; that can't even come close to what nature's doing, and it never will.

———— „ ————

Up, Up, and Away: Balloon and Helicopter Yarding

There are two main reasons why cable yarding is perceived as gentler on the environment than tractor yarding: it uses fewer roads, and it lifts the logs off the ground. Lifting multi-ton logs, however, is not an easy task. To raise them totally off the ground in a skyline system, cables have to be suspended at both ends, either from spar trees or from mobile yarders. On concave slopes this can be done, but on convex slopes it is virtually impossible to raise the cable high enough above the ground.

In the 1970s, some skyline systems started using giant helium balloons to provide lift. Logging balloons measure over 100 feet in diameter and can provide the lift for payloads weighing over 20,000 pounds. The balloon is attached to a cable system directly over the carriage that transports the logs; since the pull of the balloon is straight upward, it suspends the logs in midair, while the yarder guides both the logs and the balloon gently along the cables to the landing. The balloon eliminates the need to suspend long cables from a series of spar trees, for the balloon itself functions as a mobile spar tree, with the suspension always centered directly over the logs. The logs, therefore, can be yarded without ever touching the ground, eliminating all soil disturbance on the slopes and protecting the logs themselves from damage by rocks, debris, or irregular terrain. And since the balloon can function on cable systems up to a mile in length, the haul roads and landings are fewer and farther between.

The merits of balloon yarding are obvious, but it has its problems as well. The equipment is costly, with the helium to inflate a single balloon running over $25,000.[8] Since the balloon is difficult to control, the system is ill-adapted to any kind of selective or shelterwood cut in which residual trees might be damaged. With balloon yarding best suited to large clearcuts on steep terrain, it was used often during the 1970s and '80s, when clearcutting was still in vogue. But today, with more harvesting being done by selective and shelterwood methods and with regulatory limits on the size of clearcuts, this type of capital-intensive logging is not so appropriate. Currently, most balloon logging is done in Canada on larger tracts of land.

The ultimate logging machine is an airborne, infinitely mobile yarder: the helicopter. The image of a logging helicopter can assume mythic

An airborne yarder: logging by helicopter

proportions. No matter how tough the terrain, no matter how remote the area, the chopper can gently pluck the trees from the woods. Reaching down for a log here and a log there, the helicopter can practice selective logging while never disturbing the ground. Because they do not tamper with the earth itself, helicopters have captured the imagination of environmentalists. And they have captured the imagination of loggers as well, since they provide instant access to the innermost reaches of the forest.

Is the helicopter an unmitigated blessing? Can it solve our environmental problems while simultaneously filling our sawmills with logs? Its benefits are obvious, but what are the costs?

A brand new logging helicopter with a hefty payload capacity will cost somewhere around $15 million. And the capital expense is only the beginning. A logging helicopter capable of lifting large timber consumes 525 gallons of gasoline per hour, or over 4,000 gallons for an average working day. The crew is twice the size of a normal cable show; the helicopter requires, in addition to the usual workers on the logging site, a pilot, a copilot, alternate pilots, and a team of mechanics to service the machine while on the job. The high capital expense, immense appetite for fuel, and costly labor requirements result in operating costs of up to $4,500 per *hour*.[9]

Yet for all this money, the helicopter is not always working. Since it operates in the air, the helicopter is more vulnerable to the weather than are other logging tools. The logging season in the Northwest generally extends from April to October, yet for about 35% of that time helicopters are unable to operate because of wind, fog, or rain. Even while on the job, the helicopter must stop often for refueling and maintenance inspections. Not infrequently there are delays in locating the logs and attaching the payload to the hovering yarder, and sometimes the load must be returned because it weighs too much or is improperly attached. Altogether, the helicopter is successfully operating only about half the available time.[10]

Because of the high costs, helicopter loggers feel pressure to maximize operating time and minimize down time. There is a natural tendency to stretch the helicopter to its limit, a constant push to increase speed and productivity. If the time required to yard each turn of logs can be decreased by a mere fifteen seconds, the daily savings to the operator can amount to over $1,000.

The pressure on the workers is understandably intensified. They are dealing with expensive equipment, and precision and efficiency are at a premium. As in any work situation, there is an unending tension between production and safety. As the workers adjust their pace to the timing of the helicopter, they are tempted to take chances they might otherwise have avoided.

Perhaps the most dangerous of all logging jobs is that of the hooker in a helicopter show. In attaching the long tag lines from the helicopter to the chokers that have been set around the logs, the hooker must play with bouncing cables and dancing logs. He works under the intense rotor-wash winds of the chopper, which blow dust and debris randomly about, and which often detach limbs from nearby trees. He also suffers occasional shocks from the static electricity that builds up on the tag lines. Yet for all his trials, the hooker is the vital link in helicopter logging, the man who must produce with absolute efficiency in order to avoid costly delays. Though helicopter logging is still in its infancy, there are already several cases of hookers being killed or hospitalized.

Helicopter logging has environmental consequences that are not often recognized. As with any capital-intensive tool, more pressure is placed on maximizing the harvest in order to meet expenses. And although helicopters eliminate the need to build roads into the woods, they also require landing pads far larger than the landings used in more conventional systems. If suitable sites for landings can be found, environmental damage can be minimal; but the helicopters are often used in rugged terrain, and the construction of the landings can sometimes lead to considerable erosion in a concentrated area.

In the long run, one of the biggest problems with helicopter logging stems from the very selectivity of its operations. The helicopter picks and chooses its targets, yet since its expenses are so high, the timber removed from the forest must be valuable enough to make the operation worthwhile. What this means is that the best trees will be removed, while the inferior trees are allowed to stand. High-grading, one of the cardinal sins of forestry, is currently enjoying a renaissance, disguised as environmentally protective aerial logging. In any form, high-grading degrades the gene pool of a forest, so the net impact of helicopter logging may be the gradual deterioration of the quality of timber in succeeding generations.

Finally, there is the problem of high fuel consumption. Helicopter logging might save one small area of the earth's surface from being disturbed, but only at the expense of environmental degradation in faraway places. The 525 gallons of fossil fuel consumed by the helicopter every hour have to come from somewhere. Logging helicopters place an additional demand on existing petroleum supplies, and any additional demand these days can be met only by developing marginal sources of fossil fuel, often at great cost to environmental quality. Indirectly,

Taking the mill to the tree: the mobile sawmill

helicopter logging in the Northwest may cause or contribute to strip-mining, air pollution, and, perhaps, oil spills.

How can we justify this sort of trade-off? How can we weigh erosion in the woods against air pollution 3,000 miles away? How many tons of sediment are equal to the life of one hooker?

Helicopter logging is not "The Answer." It is costly for contractors and dangerous for workers. For environmentalists, it creates at least as many problems as it solves. Indeed, there is some question as to whether it has benefited environmental interests at all. Helicopters are used not only to log the existing commercial forests more gently, but also to reach into remote, untouched regions—potential wilderness areas, in some cases—and pick out a few choice trees. Under the guise of environmental quality, they have extended the commercial domain.

A Moveable Beast: The Portable Mill

Each of the various technologies presented here—cats, horses, cables, balloons, helicopters—involves removing trees from the forest for transport to a sawmill. Today, there is an exciting new technique which turns conventional logging on its head: instead of transporting the logs to the mill, the mill is brought to the logs. Around 1970, a couple of loggers from Oregon placed a small engine with a circular saw on a set of metal tracks mounted on a log. What the loggers had invented was a truly portable mill which, in many cases, was easier to move around than the logs themselves.

At first, these portable mills were used primarily for salvage logging. Back when good timber was far more plentiful, loggers left significant quantities of perfectly usable wood lying around on the ground. When the timber supply started to dry up, these salvage logs became more valuable, even though some of the wood had rotted while the logs lay on the ground. Starting in the 1970s, a small, decentralized army of salvage loggers combed the woods for lost logs. But why truck out the entire log when only a fraction of it was still usable? The mobile mills made salvage logging economically viable.

The early mobile mills had circular blades with large saw kerfs, generally turning $\frac{3}{16}$ of an inch of wood into sawdust for every cut. With large salvage logs, this did not matter so much. During the 1980s, however, woodsmen started using the mobile mills on small trees from selectively cut forests. Today, portable mills such as the Wood-mizer commonly use bandsaws with a $\frac{1}{16}$-inch kerf or less, making it feasible to utilize logs of any dimension. Since the mills can be dragged into the woods by an ordinary pickup truck, they bypass the need for wide roads and logging trucks. The portable mills are invaluable tools for site-specific, selective cuts where large equipment and major haul roads would be environmentally destructive and economically unsound.

Like horse logging, cable logging, and all the rest, mobile milling has its appropriate uses—but it is not a panacea. When large quantities of wood are to be milled, it is more efficient to take the logs to ordinary mills. The holistic forester will do well to use the mobile mill—or any of the other logging tools—only when circumstances call for it. Each decision has to be made as the situation arises, not on the basis of an *a priori* belief in the beauty or integrity of any single method.

Who Owns the Land?

7

INDUSTRIAL OWNERSHIP
Time Is Money

The virtues of holistic forestry seem obvious; it is really just thoughtful, sensitive stewardship of the land. It treats nature as an ally, not an adversary. It considers each site according to specific needs. It is, quite simply, forestry that cares about the future. Why, then, is holistic forestry so rarely practiced?

Often, the fate of the forests is determined by managers in distant offices who are not necessarily guided by sound silvicultural criteria. These managers live in a world driven not by sun, wind, earth, and rain but by economic and political realities. Forestry is not practiced in a social vacuum. All the scientific knowledge—and all the best intentions of on-site workers—will come to no avail unless we understand, and can change, the economic and political factors that interfere with good forest management.

There are three basic types of forest ownership in this country: public, private, and industrial. Each type has its own set of blinders, infrastructural forces that encourage short-sighted, exploitative practices while discouraging far-sighted forestry. What are these forces? How do they operate in everyday affairs?

The Corporate Mind

The timber industry owns approximately 15% of the timberland in the United States. This figure varies significantly by region, ranging from 9% in the West to 19% in the South.[1] The reason the industry owns land is obvious: to provide a source of timber and pulp for its processing plants. Although the mills will always be partially dependent upon other sources of raw material, their future is more secure to the extent that they can grow their own trees.

On the surface, it would appear that the industry should invest heavily in its growing stock. In practice, however, the timber companies spend only a small percentage of their revenues on reinvestment in the resource base. Perhaps timber is a "renewable resource," but the forest products companies are not in fact renewing it as vigorously as they could. The annual net growth of softwood trees on forest industry land is only 77% of the amount harvested; on nonindustrial private land, by contrast, the net growth of softwood trees is 127% of the annual harvest; on government land, softwood growth is 146% of the harvest.[2] The timber industry, in other words, does not seem to be providing for its long-term interests.

Don't the companies care about their future? Don't the mill owners want to maintain their resource base to provide employment for their children and grandchildren? Of course they do, but from a strictly economic point of view, it is difficult to grow and maintain real forests on their own lands. To understand why timber companies do not find it feasible to make long-term investments, we must examine the peculiar interrelationships among time, timber, and money.

When a corporation chooses to invest money in timber, it effectively chooses *not* to invest that money elsewhere. Money invested in another field will earn interest or pay dividends on a regular basis; investment in trees, on the other hand, will have to wait several decades to return a profit. When a profit is finally realized by harvesting the timber, the returns must approximate the profits that could have been made from other forms of investment. The revenue from a single crop of trees must be high enough to justify tying up capital for so many years. In other words, *part of the cost of growing trees is the interest accrued to the initial investment.*

In economic terminology, we speak of the *opportunity cost* of capital: there is always an opportunity to do something else with your money. The opportunity cost of timber is extremely high because the capital is tied up for such a long period of time. Depending on the interest which could be made in other investments (called the *guiding rate of interest,* the *hurdle rate,* or, misleadingly, the *discount rate*), the opportunity cost can become a prohibitive factor in any long-term forest investment. For every dollar initially invested, a tree that takes 80 years to mature will have to return $23 at 4% interest, $224 at 7% interest, or $2,048 at 10% interest. If the guiding rate of interest is high, investments in the future resource base become financially untenable, since they won't be able to compete with other capital investments. When the cost of interest is taken into account, there is no genuine "long term" in the practical world of business.

To demonstrate how interest rates render long-term planning financially unsound, one study calculated the *soil expectation value* (SEV) of a hectare of land that was to produce a crop of trees every hundred years. (The soil expectation value is an economic measure of the capacity of unstocked land to produce timber—adding the revenues, subtracting the costs, and accounting for interest.) Strangely, the guiding rate of interest had a far greater impact on the SEV than the actual productivity of the land. If the productivity remained constant, the land was worth $56,723 at 1% interest but only $7 at 10% interest. A loss of productivity, on the other hand, had only a minimal impact. If the soil deteriorated to the point that the volume of each succeeding crop of trees decreased by 10%, the SEV (figured at a constant 5% interest rate) declined from $741 to $740.43—a loss of only fifty-seven cents. If the land lost 100% of its productivity after the first generation of trees was harvested—if it literally fell into the ocean—the SEV would diminish by less than 1%.[3]

The implications of these figures are profound: when measured in crude dollars and cents, the future of the forest is not economically relevant. From a strictly business perspective, the long-term fertility of the soil simply doesn't matter. If a company has a chance to invest a mere one dollar per acre on soil improvement that will double the growth of the trees 200 years hence, it is economically foolish to make that investment. Unless each dollar will increase the worth of that future tree crop by tens of thousands of dollars, the company will just be pouring money down the drain.

The length of the crop rotation, like the interest rate, has a significant effect on the economics of timber. With a constant 5% guiding rate of interest, the return on a one-dollar investment will have to be $7 for a 40-year rotation, $30 for a 70-year rotation, or $131 for a 100-year rotation. Naturally, the investment goals for the shorter rotations will be easier to meet. Trees will be harvested earlier in order to avoid the large interest costs that accrue during the longer rotations.

Because of the financial incentive to shorten the cycle, the *economic maturity* of timber occurs long before the *productive maturity*. Economic maturity is the point at which a new investment would be more financially lucrative than a continuation of the original investment; productive maturity is the point at which a new crop will produce more timber than the original crop. The time of harvest is determined by the specific goal of the forest managers: Do they want to make more money, or do they want to produce more timber? These are entirely different objectives, and they lead to entirely different management schemes.

In a sense, the decision of when to harvest is not left to the timber companies; it is the marketplace that decides. Consumers want more wood, but they want it at the lowest possible price. In order to keep down the price, the companies naturally try to minimize the cost of interest. When a company harvests at economic rather than productive maturity, it is simply responding to the laws of economics—and to the wishes of consumers who want cheap wood. A company that does not respond to the market is unlikely to stay in business.

Economic maturity is of course dependent upon the guiding rate of interest; when a continued investment in timber fails to match the guiding rate, it is time for the trees to be cut. But how is productive maturity determined? Foresters have a powerful analytical tool for relating productivity with time. They calculate the average annual growth of a tree, computed over its entire life span, and call this the *mean annual increment* (MAI). The MAI is used to gauge the productive maturity of a tree: when the yearly growth falls below the MAI, it is time to cut the tree down and start over; the tree cannot meet its own standards for production. Conversely, when the annual rate of growth remains higher than the MAI, the tree should be allowed to continue growing; it is doing better than its average, and presumably better than could be expected of its replacement. In order to maximize production, foresters need only calculate the time at which the MAI starts to decline. The culmination of mean annual increment (CMAI) determines the rotation cycle which will produce the most timber.

Timber companies, however, cannot afford to wait for their trees to produce to maximum capacity. In order to turn a profit, they reap the returns from an early harvest and quickly invest in a new crop. The large annual increment in wood fiber is offset by the interest being charged to the original investment. For a typical Douglas-fir site,[4] the best economic rotation at 5% interest is to harvest every 36 years, whereas the CMAI is not reached until 64 years.[5] Economic maturity is achieved much more quickly than productive maturity. By harvesting the trees in their prime, the timber company ignores approximately three decades of peak growth, but it cuts the rotation time practically in half. Instead of continuing to pay interest on its tied-up capital, it realizes a quick profit on the first investment and moves on to the next. Some of this money will go toward replanting, while the rest can be invested elsewhere. In essence, the company gets two harvests instead of one, as well as the use of surplus capital for almost thirty years. The end result: revenues for harvesting on a 36-year rotation are approximately twice those of a 64-year rotation.

Ironically, to maximize profits a timber company has to cut corners on production. Worse yet, the wood from early harvests is distinctly inferior to the high-grade lumber fashioned from mature timber. Adolescent trees contain a disproportionate amount of soft and spongy sapwood, as well as numerous knots from the branches that have not broken off; older trees, on the other hand, can be made into clear, strong, tight-grained boards. Generally, trees from commercial species such as Douglas-fir must be a foot in diameter before they contain even a modest proportion of quality sawtimber. If Douglas-fir[6] is harvested at

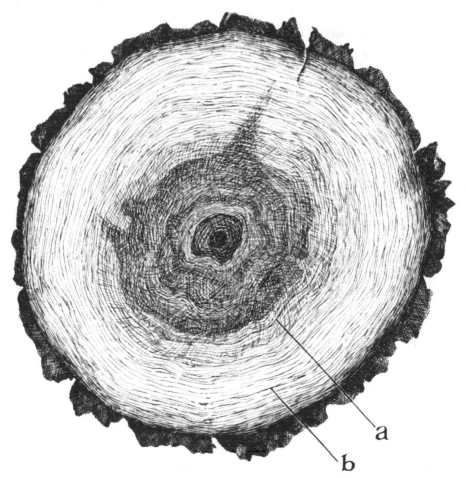

When should this tree have been cut? (a) Economic maturity; (b) productive maturity

36 years of age, the yield from 12-inch-wide or larger trees is less than 10,000 board feet per acre. At 64 years, the yield from a similar site would be about 50,000 board feet per acre.[7] A company that harvests at economic maturity will get less than 20,000 board feet per acre in 72 years (two rotations); if it were to harvest at productive maturity, it could have obtained two-and-a-half times the quality sawtimber in a shorter period of time.

The implications of this discrepancy are serious. It is well known, of course, that the economic incentives of private industry do not always coincide with the public interest. We accept the fact that the private sector must sometimes be required by legislation to take actions that are economic liabilities: they must be made to clean up their own wastes, for instance, or to provide safeguards to the consumer. The problem here is even more basic: the timber itself is sacrificed for the sake of profit. The strongest arguments in favor of private enterprise are based upon efficiency and production: corporations may not always act according to the best interests of the environment, but at least they get the job done, they deliver the goods. The private sector, we assume, produces what we want to consume. Not so in this case. The peculiar relationship between time and timber causes private industry to fail at its ostensive task: maximizing production.

The problem is not with the companies themselves, but with an economic system in which interest rates are pitted against the time it takes to grow trees. Private enterprise, operating according to economic necessity, is simply not suited to the job of producing the most and the best timber products. When the guiding rate of interest exceeds 3%, as it does in the current economic landscape, sound economic practices on the part of timber companies are literally counterproductive. Quality sawtimber cannot be produced on corporate lands, except at exorbitant prices that offset the cost of interest—and which the consumers, at least so far, are unwilling to pay.

The same economic reasoning that favors shorter rotations causes the timber companies to shorten the natural cycle of forest succession. By bypassing the pioneer stage, they also bypass many years of interest. And in their choice of methods for eliminating unwanted brush, any savings they can make will be greatly enhanced by the guiding rate of interest, since they can take the money saved and invest it elsewhere. Area-wide treatments such as the spraying of herbicides are preferred to the more personalized (and generally more expensive) site-specific treatments such as manual release. Any extra input during the early years of the cycle must produce a much-magnified output, or it simply isn't worth the money. If the application of herbicides is $10 per acre cheaper than hand-clearing, the savings will amount to several hundred dollars per acre by the time the tree is finally harvested. If the companies don't think that the end product from hand-clearing will be several hundred dollars more valuable than the end product from spraying with herbicides, then they do not feel justified in making that extra $10 investment.

As the guiding rate of interest gets larger, silvicultural decisions become increasingly more dependent on economic criteria. At 1% interest, every dollar saved now will result in a $2 savings seventy years down the line; at 10% interest, every dollar will result in a $790 savings. Although these are the extremes, the fluctuation in the guiding rate of interest enters significantly into real-life decisions. Forest economists speak of the *present net worth* (PNW) for a given site over a defined planning period—the sum of the revenues minus the sum of the costs, taking the guiding rate of interest into account on both sides of the balance sheet. If the PNW falls below zero, the project becomes an economic liability; if the PNW remains positive, the project is worth undertaking. But the PNW hinges upon the guiding rate of interest. At a 3.5% rate of real interest, for instance, manual release and precommercial thinning on the Hoopa Valley Indian Reservation lead to a positive PNW; at 4% interest, these same projects generate negative values and would lose money for the tribe.[8] Similarly, interplanting Douglas-fir stands with green manure trees such as red alder might make economic sense at low rates of interest, but at high rates this investment in the future forest cannot be justified.[9]

How can the magical guiding rate of interest, or "discount rate," be determined? This is no easy task. Basically, it is no more than the projected rate of real interest that is expected to prevail throughout the economy in the years covered by the planning period. Needless to say, the exact rate of interest forty years hence is anybody's guess. This uncertainty makes economic planning exceptionally difficult. Timber company managers, in order to avoid being devastated by high interest rates in the future, must make their projections reasonably high; they are safer if they assume the worst. But the assumption of a high rate of interest both limits investment and increases the importance of time as a determining factor in management decisions. Less money can be spent on the future, while rotation cycles become even shorter.

Why, one might ask, would *anyone* want to invest in timber? If the investment is so sensitive to the guiding rate of interest, and if the interest rates of the future are so hard to predict, isn't it just too risky?

If timber had no economic value prior to harvest, the risks would indeed be too great. In fact, however, timber is traded on the open market long before harvest; it has economic value even as it grows. Timberland buyers and sellers are speculating in a future product. The speculation can be worthwhile because the investment appreciates in three distinct ways. (1) With each passing year, a tree grows upward and outward. The annual increase in volume varies by species and site, but it is generally of significant magnitude for several decades. (2) As the tree matures, its end product takes on greater value. At first it can be turned only into pulp, then into low-grade lumber, and finally into high-grade lumber or veneer. This change is called *ingrowth*. (3) As available resources become more scarce, price increases tend to outstrip the overall rate of inflation. With timber values increasing in three ways simultaneously, timber owners can realize healthy and competitive profits. From the 1960s through the 1980s, nominal returns on timber investments were approximately 12%; real returns ranged from about 5% to 8%.[10]

These last two factors—ingrowth and price increase—encourage timber owners to wait rather than cut, serving as partial checks against early harvesting. The effects of high interest rates, however, are potentially more significant than the increased price of wood products. It is hard to imagine, for instance, that the real price of lumber will be 131 times higher in a hundred years than it is today, although that in fact would be the effect of a 5% real interest calculated over a century.

The ultimate test of profitability for timber owners, as for any capitalist enterprise, is the *internal rate of return* (IRR): the compounded annual interest rate earned on the initial investment. If the IRR compares favorably with the guiding rate of interest—what capital could generate if put to some other use—then the project is worthwhile. Today, growing trees can produce a respectable IRR and is therefore a good investment—but only with short rotations. The shorter the cycle, the more predictable the results. Frequent harvests generate capital for repeated investments, whether in timber or in some other field. Just as second-growth trees are more manageable than old-growth, so too are investments that last only 30 or 40 years easier to control than those that take twice as long to turn a profit.

In order to shorten the rotations as much as possible, investors are increasingly moving toward producing pulp instead of sawlogs. The pulp can then be pressed together, simulating old-fashioned lumber. Without an understanding of economics, we might suspect that it makes more sense to produce real boards than to glue wood pulp. But pulp can be grown much more quickly, bypassing the incredible impact of time on forest investments. Pulp plantations produce marketable merchandise in a small fraction of the time it takes to grow real timber. From an economic standpoint, the time frame for growing pulp—say 15 to 30 years—can be handled within a capitalist economy; the time frame for regenerating a real forest—say 70 to 300 years—is incompatible with capital investments that must produce competitive rates of return.

Time, in the terms of forest economics, is measured in years or decades but never in centuries. With no economic incentive to plan beyond the next crop or two, investments in soil structure or erosion control cannot be justified financially. Any notion of spending money to repair a damaged ecosystem is ludicrous from a business point of view. Environmentalists claim that the timber companies are acting unethically by ignoring the distant future, but the problem is actually fiscal, not moral. The problem is created by a system in which we all play a part, consumers and producers alike. Given the fact that a corporation is an economic entity, why should it invest in activities that show no financial reward? Perhaps it will make token gestures, but these amount to no more than charitable contributions or affirmations of good will. There is no *structural* reason for a corporation to practice the kind of forestry that will lead to a healthy, productive stand of trees 200 years from now.

The whole economic edifice is entirely rational—but it is based on a logic that has nothing at all to do with silviculture or ecosystem management. Financial reasoning leads the companies to cut trees more frequently than they should, lessening the total production of quality sawtimber. It leads them to ignore the principles of forest succession that should form the basis of sound forest management. It leads them to harvest timber from areas that are too sensitive to withstand the onslaught of heavy equipment, too steep to avoid subsequent erosion, or too exposed to generate a new crop of trees. It leads them to pay little heed to nontimber values such as water quality, fisheries, and wildlife habitat. And it leads them to skimp on investments that would benefit tomorrow's timber, since the nature of interest rates renders long-term, slow-return expenditures fiscally unwise.

Corporations at the close of the twentieth century, however, do not operate exclusively according to economic principles. Increasingly, they function as public entities that are legally responsible in some respects to furthering the good of society. Whether willingly or not, they are subject to regulatory constraints which tell them to act against their immediate economic self-interests. The purpose of regulations is to account for "externalities"—factors which do not show up on the balance sheet.

Ironically, the preservation of the resource base for the distant future constitutes such an externality. Whether or not the future productivity of a site is economically significant, timber companies must preserve the integrity of the soil in order to satisfy legal requirements. If reforestation expenses were treated as discretionary investments, they would be hard to justify financially; but by defining reforestation as a necessary expense charged to the previous harvest, the regulatory agencies are able to make sure that trees get planted—even though tree planting might not be profitable if interest rates are taken into account. In a sense, these regulatory restraints help the companies think about the future, since they have little economic incentive to do so on their own.

Some timber companies go a step further: They take noneconomic factors into account *voluntarily*, not just because they are forced to comply with the laws. In particular, family-controlled "dynastic" companies are more likely to take future productivity into consideration, even though there is no profit to be made by doing so. While the impact of interest favors short-term investments, "dynasties" sometimes view time more leniently. A healthy future for the forest means jobs for the children and grandchildren. Although employment opportunities for unborn offspring do not appear on the balance sheet, family or community-oriented businesses operate as if this type of human variable has value. Harvesting timber, to some executives, represents more than just a way to get rich; it's a way of life worthy of being preserved.

Company Town

Before 1986, the shining light of the western timber industry was the Pacific Lumber Company, which operated not only the largest redwood mill in the world but also one of the last company towns in the United States: Scotia, California. Founded in 1869 by the Murphy clan from Maine, Pacific Lumber owned almost 200,000 acres of prime timberland, including the only significant tracts of old-growth redwoods not in government preserves. (In fact, much of the world-famous Humboldt Redwoods State Park was carved out of former Pacific Lumber holdings.) Historically, the company had no difficulty in keeping its mills supplied with trees from its own lands; unlike most lumber processors, it did not have to rely on purchasing timber from the government or other private owners. With timber to spare, its rate of harvesting was modest. By all accounts, the Pacific Lumber Company (often called PL or PALCO) managed its lands on a sustained-yield basis long before many modern-day environmentalists were even born. Increasingly, company foresters practiced modified selective cutting in preference to clearcutting. Pacific Lumber, in short, was living proof that corporate ownership and good forestry were not mutually exclusive.

On a more personal scale, PL coddled its workers like no other company. After a serious strike in 1945 that shut down the mill for several months, the company decided to combat unionization with a simple but effective strategy: offer the workers just a little more than the unions and union shops could afford to match. The strategy worked, as evidenced by the lowest worker turnover rate in the industry. Employees also received yearly bonuses of up to 7% of their salaries. To reward the faithful, five-year men were granted seven paid holidays and four-week vacations with pay. Scholarships were offered to college students coming from company families. Workers were covered by medical insurance and a liberal pension plan in which the company doubled the employees' contributions. And most important of all, the company offered job security in a troubled world—not only for the present workers, but for generations to come. Because of its sustained yield harvesting from company-owned lands, Pacific Lumber expected its supply of timber to last forever. In an era of shrinking supply, layoffs, and mill closures, the promise of future employment was the greatest possible benefit the company could offer the residents of Scotia and nearby Rio Dell.

Today, that kind of security is gone. The problem, it seems, was that Pacific Lumber had managed its lands too well. In 1986, according to a survey of its holdings conducted by an independent consulting firm, the value of its stock should have been somewhere between $60 and $77 per share. The actual price on the New York Stock Exchange, however, hovered around $28 per share. Pacific Lumber was ripe for a takeover.

Charles Hurwitz, a corporate raider from Houston, began quietly accumulating shares. So too did various business acquaintances, among them Ivan Boesky and Boyd Jeffries. Gradually, the price of a share crept up to $33. Then on September 30, 1986, Hurwitz announced his takeover bid: he would pay $36 for any and all shares of the Pacific Lumber Company. The bid was later raised to $38.50, and then to $40. Finally, Pacific Lumber's Board of Directors agreed to accept the offer. The company—the land, the trees, the mill, and the town itself—was no longer in the hands of the Murphy family and friends. The people in Scotia, California, would henceforth have their lives and their livelihoods determined from afar.

To finance the purchase, MAXXAM (Hurwitz's group of acquired companies and holding shells) needed to borrow close to a billion dollars. About half of the financing came in the form of junk bonds arranged by Michael Milken of Drexel Burnham Lambert. Since that time, Milken, Boesky, and Jeffries have been implicated in Wall Street's largest insider trading scandals. Hurwitz, however, appears to have emerged unscathed, in firm command of the world's largest producer of redwood lumber. Hurwitz made a personal appearance in Scotia shortly after the takeover, telling the assembled workers, "You know what they say about the golden rule: He who has the gold rules."[11]

But how would MAXXAM pay off its debt? Would the redwood trees still be cut at the same slow, steady pace which insured a sustained yield and future employment for the residents of Scotia?

Bill Bertain

Attorney for Former Pacific Lumber Shareholders

"My grandfather, my mom's dad, came here to Humboldt County in 1882. My other grandfather, Louis Bertain, came in 1893. At first both of my grandfathers worked in the woods. Then my mom's dad got a farm, and grandpa Louis got into the laundry business here in Eureka.

"In about 1918 the state legislature passed a law saying that any lumber company having a bunkhouse had to provide one clean sheet a week to each of the men. So Pacific Lumber started using my grandfather's services, because they had hundreds of guys working in the woods (probably over a thousand including the mill) who each needed a clean sheet a week. They'd bring the sheets into Eureka by train and my grandpa and my dad and my uncles would wash the stuff and ship it back. Finally the company superintendent suggested, 'Why don't we just move you out to Scotia?' So that's what they did. In 1920 they put all the equipment on a train and brought it down to Scotia and moved it into a place on Williams Street called 'The Green Goose,' which up to that time had been a house of ill repute. I guess they had decided to clean up the town—with clean sheets too.

"My dad called Scotia 'home' from 1920 until he died in 1988. He lived in company housing from 1920 until 1946. I was the youngest of ten kids. I was born in Scotia, as were my nine older brothers and sisters. It was a great place to grow up. The school was good, good ballpark, skating rink, swimming hole with three diving boards. Later, they built a gym with an Olympic-size pool. You

The log pond at Scotia: balanced on board feet

could wander all over the town and the hills. And the company provided for the employees like no other company in the county ever
did, like very few in the whole country. In the thirties work was
slow, but the company kept people on the job. They waived the
rent, or at least lowered it. Scotia was practically insulated from the
Depression. Then again in the early eighties, with the big timber
recession, they cut people back to four days a week rather than lay

anybody off. You didn't have to worry too much. You had job security—not just for yourself, but for your kids. At least that was the
conception. In 1984, most kids graduating from high school had a
reasonable expectation that they could work in Scotia until they
retired at age sixty-five, and probably that their children and grandchildren could work there too. That was because of the tremendous
resource they had, and because of the philosophy of the company
toward both the land and the people.

"In 1969 they had their centennial banquet, celebrating 100
years of Pacific Lumber. Mal Coombs (he was quite a presence)
sang a ballad that ended with something like: 'I believe in God and
PL.' That got a big roar. Lots of fun. Then Stan Murphy got up and
said, 'I just want to dispel any rumors you may have heard that we
will sell Scotia. I want to make sure that everybody here understands . . . WE WILL NEVER SELL SCOTIA!' That brought the
house down.

"Then in August 1972 Stan died, which I think was almost a pre-
requisite for the takeover. He wasn't very old, only 51 or 52. I was
working in the laundry at the time, and my dad came out to tell me
the news. He was crying, and then he told me something that was
very significant: 'That's the worst thing that's happened in Humboldt County in 40 years. It's worse than the '64 flood.' As it
turned out, I think he was right, because Stan probably wouldn't
have stood by and watched this thing happen. Without him, the
company was less prepared to deal with it.

"I grew up knowing the Murphys. Our families had kind of a
mutual friendship. I taught the boys, Woody and Warren, how to
serve mass. In March of '85, I started doing some legal work for
Woody, which continued through the summer. When Hurwitz
announced his takeover bid, Woody asked me to keep an eye on
this, because we might have to fight it. And I've been fighting it
ever since.

"In 1981, the shareholders of PL had decided that if anyone held
over 5% of the shares, any hostile takeover would require an 80%
approval. That would have been hard to get, so MAXXAM, when it
was getting ready for the takeover bid, had to stop buying at 4.9%.
Instead, we allege that they got Drexel Burnham to buy shares for
them. Drexel Burnham filed an affidavit saying they hadn't bought
shares for MAXXAM, but Ivan Boesky has now admitted that he
had bought shares for Drexel. Also, if you hold more than $15 mil-
lion worth of shares, you are required by the Hart-Scott-Rodino

Act to give notice to the target company what your purposes are, and you're supposed to stop buying shares for a period of time. That gives the target company an opportunity to defend against a takeover. So Hurwitz stopped at $14.5 million, but he had Boyd Jeffries buy another $14 million. 'Parking stocks' like this is illegal.

"Salomon Brothers, which had been advising the old Board of Directors, had estimated that the value of each share was between $60 and $77. The final tender offer, as you may recall, was only $40. But the Salomon Brothers' estimate was not disclosed until after the merger agreement. Had it been disclosed, people would not have sold out for only $40 a share. It would have changed the whole terrain over which the battle was being fought.

"Why did the Board of Directors agree to sell to Hurwitz? That's been a big question all the way through this thing. The date of the merger agreement was October 22. On October 18 Hurwitz filed a suit against the Board of Directors in the state of Maine, where Pacific Lumber had been incorporated, charging that they were not acting in the best interests of the stockholders if they refused the offer. That scared the hell out of the Board, and on the 19th or 20th they started negotiations. I think the Board would have defeated the suits. Even if they had lost, they would have paid out $20 or $30 million bucks to save a way of life and a whole economy. It would have been worth it. Instead, they caved in. They got outsmarted. They had lousy advice and they panicked.

"The Board was asleep at the switch. They hadn't done a complete timber cruise since 1956, so they didn't even know how much they had. As it turned out, Pacific Lumber owned at least $2 billion worth of timber and timberland, probably more like $2.5 or $3 billion—and that's not counting the other assets like the mills and the other buildings in Scotia and the welding division. And Hurwitz paid only $870 million for the whole company!

"It was an incredible bargain, but Hurwitz still had a $660 million junk bond debt, plus a bank debt and various fees. His total debt was about $1 billion. Immediately, as soon as he acquired the company, he sold off the welding division for $325 million. He doubled the harvesting rate, maybe even tripled it. He hired 300 more employees and increased the number of hours, overtime, and extra days. That's had some negative impact on the social fabric. People are working more hours, but they don't have any more time with their families. Last summer, he started having the guys work earlier in the morning. He was trying to not use electricity during the peak afternoon hours so they could sell it to PG&E for a greater price. Can you imagine what that does to family life? How can you go see your kid play Little League when you have to get up at 3:30 or 4:00 in the morning?

"Then there's the pension fund, which had built up a huge excess. Their obligations required only $40 or $45 million, but they had at least $96 million in it. Everybody knew this, so on October 9 the Board of Directors amended the rules to keep the funds out of the reach of corporate raiders. They stated that in the event of a hostile takeover, any excess in the pension fund shall immediately invest in the existing employees and retirees. I thought that was pretty clever, but Hurwitz challenged this in his October 18 suit. Unfortunately, it was negated when Hurwitz got an agreement from the Board of Directors—it was no longer a 'hostile' takeover. So Hurwitz just *took* the surplus from the pension fund to pay his debt. He also bought a cheaper annuity plan from Executive Life with no cost of living increase, and it isn't federally guaranteed. It's a real bargain basement plan, the cheapest and the worst on the market. Executive Life just happens to have purchased large quantities of Hurwitz's junk bonds. Is that a payback?

"This whole thing has been a real tragedy, but it's hard to fight against MAXXAM. They have a lot of power, and they're not afraid to use it. In January of '92, my brother had his laundry booted out of Scotia. Campbell and Hurwitz knew that I was doing everything that I could to rectify the situation and seek justice, and they didn't like it. I'm not 100% sure, but most people that know anything about what's going on feel that they were getting rid of my brother because of me.

"Most people in town are still pretty intimidated. There's a lot of fear. You can lose your job if you complain too much. It's a real sad situation. The people know that the faster they work, the sooner they'll be out of a job. They used to be able to feel that they could work there the rest of their working lives, and now that security is gone. They realize exactly what's going on.

"Some people, on the other hand, have developed some degree of nastiness toward environmentalists. That's understandable, but lots of their anger is directed in the wrong direction. If Hurwitz hadn't accelerated the cut so blatantly, few people would have ever known about the 'Headwaters Forest.' [The controversial old-growth forest owned by PL.]

"The people who feel that Campbell and Hurwitz are the good guys argue the property rights line. I believe in property rights too, but I believe you have to acquire property by fair means, not by fraudulent means. And I think you have a duty as a steward of the land to do it right and think about future generations.

"To be a free man, you're going to have to fight. It's been a real strain financially. I'm living on borrowed money. The fight to preserve Pacific Lumber has taken eighty, ninety percent of my time for seven years. I'll probably get it back when we win the case. With Drexel, it's just a matter of divvying up the money to satisfy all these cases where they've committed fraud. But ours seems to be the biggest case of any of them. It would be great if we could return Pacific Lumber to its rightful owners. But now, since Hurwitz's refinancing in March of '93, the most likely way our case could return PL to the old shareholders is to win such a large settlement that Hurwitz would be forced to offer the company as payment.

——— " ———

Big Business, Big Government

Time and money always loom large, but leveraged (debt-financed) buyouts magnify the impact of corporate economics on forest practices. At the risk of losing the company, the high-interest debts must be paid. Increased cutting is therefore *required* in order to stay in business, not merely *preferred* in order to secure higher profits. The stakes of the game are greater.

By contrast, the economic forces that impact forest practices were relatively insignificant at Pacific Lumber before 1986. With old growth to spare, PL didn't have to worry about shortening its rotations. With plenty of timber to feed its mills, it didn't have to take more from the land than could be returned. Even the stockholders were satisfied. Because the stock was priced at only a fraction of its real value, dividends looked respectable despite the fact that they could have been much higher. A two dollar dividend represents over 7% of a $28 investment, while it would have constituted less than a 3% return if the stock had been selling at its true worth of $70.

In its own defense, the new Pacific Lumber management claims that the more recent inventories justify its increased harvest levels. Since they have more timber than they thought they had, they can afford to remove the excess without ill effect. Their aim is to cut at the higher rates for about 20 years, paying off the debt and placing the company on secure financial footing. After that, they will retreat to the pre-1986 levels on a sustained-yield basis.

The only obstacle to this plan, they say, is that they no longer have access to their own land. Legal challenges to its timber harvest plans, based on the alleged danger to wildlife species such as the spotted owl and the marbled murrelet, have severely restricted Pacific Lumber's ability to harvest old-growth redwood timber. The economic impact on the company is immense; according to company forester Robert Stephens, it could cost them as much as one billion dollars. Far from being the villain, Stephens feels the timber giant has been victimized by the environmentalists. In the words of PL attorney Frank Bacik, "It could be that for the good of the murrelet, PALCO is going to go out of business." If their remaining old growth is effectively removed from the rolls, the company certainly will not be able to maintain the current level of cutting; perhaps it couldn't even continue at its old rate and still wind up with a sustained yield.

In the old days, if the government wanted to allocate private land for public use, it had to pay for the land first. Not so today, claim company representatives from Pacific Lumber and elsewhere. When regulations keep the owners from cutting their own timber, the government essentially takes the land for free. Often, the land was acquired before the laws inhibiting its use were in effect; at the time of purchase, the buyers had a reasonable expectation that they could harvest their trees. Having paid a fair market price for the timber, they feel they should be permitted to harvest it. If they are prevented from doing so, their property is being deprived of its value by legal fiat—and the owners should be compensated accordingly.

In response, environmentalists argue that the regulations don't take value away from the property, since the value was not really there at the beginning. There never was a law stating you can do whatever you want with your land, regardless of the consequences. A company that acquired land with the expectation that it could cut down everything in sight was not paying sufficient attention to political realities, let alone environmental quality. It is not being deprived of some inherent right; it just made a bad management decision.

Should the companies be compensated for land which has been regulated to the point that it is no longer financially profitable? According to various court decisions passed down over recent years, owners deserve compensation if land is taken for the "public good" (recreational parks, scenic views, spiritual values), but not if it is taken to prevent "public harm" (landslides, damage to water quality, loss of wildlife habitat). In the case of old-growth redwoods, however, these variables are not so easy to distinguish from each other. Another factor in court decisions (and one which is easier to measure) is the extent to which regulations affect the value of the land. If over 90% of the assets are effectively tied up by regulations, even land taken to prevent "public harm" might be eligible for compensation.

In 1993, PALCO made creative use of these distinctions by restructuring the company into separate components: the second-growth stands would be held by Scotia-Pacific, while the virgin old growth would remain in the hands of Pacific Lumber. If the government decides to restrict the harvest of old growth in order to preserve the habitat for threatened or endangered species, that will impact the overwhelming proportion of the new Pacific Lumber holdings. Environmental set-asides would therefore constitute a "takings," meaning that the company would receive compensation. The second-growth timber held by Scotia-Pacific, meanwhile, will not be subject to the litigation which has been threatening the financial security of the company. Because of the legal challenges to old-growth timber, PL has not been considered a very secure investment; now, Scotia-Pacific presents less of a risk—and can therefore borrow money at will.

The story of the Pacific Lumber takeover illustrates the two extremes of the timber industry: first, the family-run, community-based mill town that carefully preserved its resource base, followed by the leveraged, debt-driven corporation which was forced, by the basic economics of timber, to institute hastier forest practices—increased cutting of old growth, shorter rotations for young growth. But Pacific Lumber is not alone. Champion International acquired St. Regis Paper and then sold several of its assets in the Northwest to pay the bill. Sir James Goldsmith acquired Diamond International and Crown Zellerbach and then proceeded to sell off several of the mills. Increasingly, timberlands are traded like cattle on the open market, subject to the dictates of national and international finance which have little or nothing to do with forest ecosystems.

In California alone, six sales accounted for over one million acres of timberland changing hands within a five-year period. The buyers in each case were corporations, and four of the sales were leveraged to the point that the buyers expected to sell off some of their recently acquired assets in order to pay off the debt. According to an official government report which took note of these sales, "There can be no doubt that heavily leveraged transactions, by design, emphasize short-term results. This can mean rapid liquidation of merchantable growing stock and lower levels of investment in management activities."[12]

The difference between a family operation and a public corporation is simple but profound: in the former, economic factors are important but not necessarily all-consuming; in the later, all decisions are made according to financial criteria. Structurally, the family company is responsible to individual owners with specific personal interests, which can (but don't have to) include such amorphous variables as employee satisfaction or community well-being; the public corporation, on the other hand, is responsible only to anonymous stockholders whose expressed goal is to make money. Since stocks can always be traded, the corporation is always for sale. Family values have no place on the New York Stock Exchange. Pacific Lumber lost its autonomy not when Charles Hurwitz made his takeover bid in 1986, but when it first issued stock that could be traded on the open market.

The key impediment to long-term management, we may recall, is the guiding rate of interest. Since capital tied up in timber cannot be invested elsewhere, the returns on forest investments must be competitive with those in other fields. For corporations, therefore, there is no way to get around the impact of interest rates on forest management. Family operations, on the other hand, are not necessarily as driven by interest rates, unless they happen to be in debt. If there is no intention of transferring the investment somewhere else, the guiding rate of interest ceases to be the all-important factor. It's the *mobility* of capital that forces timber owners to keep apace; for timbering families with no desire to change their way of life, interest rates are not particularly relevant unless they want to attract more capital. All they need is enough money to pay their employees and their bills, with a little left over for themselves.

Family Values: Setting a Good Example

The Collins Pine Company in northeastern California was founded by timber magnate E. S. Collins, who came west from Minnesota in the early 1900s. Today, Collins Pine gets its supply of timber primarily from the 91,000-acre Collins Almanor Forest—but it does not actually own the forest. In his will, E. S. Collins left the processing facilities to his heirs but the land to the Methodist Church. The timber to be cut from the Almanor Forest is determined not by the needs of the mill but by Barry Ford and three other professional foresters who manage the land. According to Ford, "My charge is to practice sustainable, selective forestry. It's that simple."

In 1941, the Collins land held about 1.5 billion board feet of timber. Since that time, 1.7 billion board feet have been removed, yet the forest still has approximately the same volume as it started with. The Almanor Forest is often toured by foresters in search of the secrets of sustainability. According to Ford, sustained yield is not a very difficult concept: just take an inventory every ten years, then proceed to harvest no more than the forest has been able to grow in the previous decade. The interest is harvested, while the principal is left intact. Unlike most industrial foresters, Ford does not appear to be in much of a hurry. He sees brush as part of natural succession, not simply as weeds that must be removed to get on with the business of planting conifers. "You've just got to be patient," he says. "You'll get trees here eventually. A hundred years for this stand is nothing."[13]

Apparently, it is still possible in this day and age for a timber company to stay in business without diminishing its resource base. But if it can happen in some cases, why can't we change the rules of the game so it can happen everywhere? Aren't there some ways in which we can lessen the destructive impact that interest rates seem to have on our future forests?

In the wake of the MAXXAM takeover of Pacific Lumber, California State Senator Barry Keene, representing the district which includes Scotia, introduced a bill limiting debt-ridden buy-outs of timberland. If purchases were not heavily leveraged, Keene reasoned, there would be less immediate need to liquidate the growing stock. Under pressure from the industry, Keene's bill died in committee. Timber companies did not want to see their investment flexibility limited by legislative control.

More recently, the California Board of Forestry has been considering ways of prohibiting short rotations. On site class I, for instance, plantations could not be harvested until they are 50 years old; on poorer sites, the minimum time would be longer. In no case would harvesting be allowed without a sustained-yield plan.

Perhaps such restrictions are appropriate, but economic forces are not easily whisked away by administrative rulings. "Sustained yield" is a relative term; growing pulp, for instance, can be sustained on very short cycles. If the definition of sustained yield is left up to the owner, almost any rotation length would qualify. If, on the other hand, a specific time frame is established, investments in timber that are profitable at shorter rotations might no longer be financially feasible. Since longer cutting cycles lead to diminished profits, growing timber on some sites might no longer be competitive with other investments. Some land would be taken out of production, causing a decrease in total available timber and a corresponding increase in the market price of wood products.

If *all* timber owners used longer rotations, more timber would eventually be produced on the land still allocated to growing trees commercially. Since harvesting at CMAI generates a much higher yield than harvesting according to economic criteria, consumers would wind up with more lumber in the long run. But if only *some* timber owners are required to use longer rotations while others are not forced to do so, those adhering to the new laws might be driven out of business by their competitors. Indeed, this has been a real problem in recent years. Since regulatory constraints in California tend to be stricter than those in other states, California timber growers claim they must operate at a competitive disadvantage. The costs of legal compliance in Oregon and Washington are less than half of those in California.[14] Investors are understandably wary of allocating capital to California timberland— and if they do, they will have correspondingly less money available to devote to the future forests.

Perhaps the only way to deal on a structural level with the economic impediments to sustainable forestry is to change the ownership of the resource. Corporations of stockholders, as we have seen, are driven by forces antithetical to long-term ecosystem management. To make matters worse, the timber industry, by definition, is focused primarily on wood processing; the function of forest ownership is only to provide a supply of raw material. Harvesting decisions are therefore strongly influenced not only by interest rates but by mill capacities. Large, capital-intensive mills in particular must operate at or near full capacity in order to turn a profit; timber is therefore cut not necessarily when it is ready, but when it is needed by the production plant.

Are processing companies with short-term interests the most appropriate owners of the timber resource? Probably not. Although the more personalized, family-styled companies might be on the wane, perhaps there are other types of nongovernmental owners that would be better stewards of the land.

Jim Rinehart

Investment Broker

"Forestry is a second career for me. My undergraduate degree is in zoology, but I went on to business school for an MBA. I spent about ten years in the retail business—which was fine for a time, but I ultimately reached the point where it just wasn't to be doing anything for me. I've always spent a lot of time outdoors; when I wasn't out there, I wanted to be out there. I did some research and found out that the forest industry was a place that I could apply my business background, spend more time outdoors, and hopefully contribute to something worthwhile. So I went back to school in forestry.

"When I made this switch, it was with the intention of working in the woods; at least that was the image I had in mind. But it didn't take me long to realize that the guy working on the ground is probably doing the most important job but has the least influence. A forester may go out there believing absolutely in what he or she is doing, spend two years putting together a great hundred-year management plan, only to find that somebody on the forty-fourth floor in some big office building in a far distant city has just sold the company to someone with a wholly different view. Everything goes out the window. I wanted to be in a capacity where I had some influence over the course of events, and that's generally in the financial end of the business.

"At first I was working with small nonindustrial landowners. But once again it became a matter of influence. I still have a deep interest in small landowners, but they're widely diverse. It's like trying to get twenty-five school kids headed in the same direction at the same time. Not that they behave like children, but they do have extremely different views of where they want to go and how. So I made a one-eighty and asked, 'What about the big landowner?'

"The first question I asked was: 'How does a forest company view its resource? How do they calculate return?' It became apparent that timber companies in general (with some notable exceptions) were primarily in the processing business. Their main interest was in manufacturing a finished product of some kind— lumber or paper. Timberland ownership was a means to an end, a way to insure a supply of raw material. But the asset they owned was long term, where their primary interest was in the short term. So there was an incongruity. I began to think that there might be other types of forest owners who would have a more appropriate time frame, such as pension funds or other types of institutional investors.

"I went to work for John Hancock, the insurance company, as a forest economist, which means I looked at what was going on in the industry with respect to supply and demand and attempted to guess what prices were going to be. That both guided their forestland acquisition efforts and enabled them to report to their clients about expected performance. On a more applied level, I devised systems for managing investment portfolios and determining portfolio strategies—deciding what properties to buy and where. I do that now, but for other investors.

"There are five primary investment managers in this business, Hancock being the largest. They all raise money from pension funds, use that money to buy timberland, and sell timber to the forest products industry. The strategy so far has been to acquire timberland that is no longer strategic to the company. Now, we're trying to take that a step further: Let's deal with strategic property, and let's do it by making direct arrangements with the processing company. Airlines don't own airplanes, and railroads tend not to own boxcars, so why should timber companies own the timberland if there is some other way to insure supply? Why should they have all that capital tied up in land when they can better use it for more immediate needs?

"Let's suppose a timber company with an established land base has an immediate need for capital—perhaps to upgrade a mill, or to diversify into another related business. An institutional investor with a long-term view can provide that capital by acquiring a portion of the company's forestland. Through a long-term agreement to sell all timber harvested from that land back to the company, the company has the same access to the raw material as though they still owned it, but with two important differences. First, they must pay full market price for the timber they buy, which provides a return to the investor. Second, they are subject to the management constraints imposed by the new owner. To the extent that the new owners have a longer-term investment horizon and no investment in the mill, the forestland itself is more likely to be managed for the long term. By placing the long-term resource in the hands of a long-term investor, we increase the propensity for better land management.

"From the standpoint of market stability, wood processors are not always the best owners of timberland. If a timber company is obliged to keep the wood flowing to and from the mill in order to stay in business, they may be forced to harvest at the most inappropriate time. I'll use a simplistic example. Lumber and log prices tend to be quite volatile, characteristic of commodities. If prices were to drop by 25%, let's say, when the mill is operating at break-even, then production must be increased by 25% to generate the same revenue. So harvest increases at the very time when it should be held back, allowing the timber to grow until prices rise again. The company cuts into capital and it becomes more difficult for the land to recover. And by putting more product into the market, they continue the cycle of low prices.

"Institutional investors can better accommodate market uncertainties. In forestry you have to wait, and pension funds are in a better position to wait than wood processors. Pension funds tend to make all cash transactions; they don't like debt. That gives them a lot of flexibility. They don't have big payments to make, so they can afford to wait out the market. They're not subject to this inexorable debt-service that must happen regardless of what the land is producing.

"If this separation of resource from processing were to occur universally so that *all* of the forest resources were owned separately by long-term investors, forestland owners in general would forego harvest in down periods, which would keep the market in better balance. The processing industry would benefit as well. Since everyone would be subject to the same constraints, higher prices could be passed along to the consumer—and that's who should be paying in the case of a supply shortfall.

"I think timber companies are now more open to these ideas because of the higher carrying costs of timberland and the need to convert to new small-log technology. The need for capital is higher, and so is the cost of hanging onto timberland. In part as a result of the environmental movement, it is now more expensive to own and manage timberland. The uncertain regulatory environment means that the harvest levels anticipated when the management plan was written may not be achievable five years down the line, which diminishes projected return. Capital is scarce, supply is scarce, and new ownership structures will evolve as a result. So there's a greater propensity in today's business environment for this kind of transaction. That seems to be borne out by the positive response we're getting from the timber companies.

"Of course the buyers are also affected by increasing regulatory constraints. We deal with this by assuming the worst, and that affects how much can be paid for the property. Which, of course, makes it relatively more difficult to agree on a price. One way to resolve this problem is to put something in the structure that says, 'If it goes the other way and things turn out better than we anticipate, then the seller can participate in the higher returns.' We write that into the contract as an incentive for the seller. If times are good, they will get a little something extra. It creates a new relationship; you become partners rather than antagonists.

"I believe in incremental progress. We're still working within a recognized economic context, but we're inserting a new layer of ownership between the processing industry and the forest resource. Twenty years from now, I suspect you'll see substantial changes in the structure of forest ownership and a marked improvement in the quality of forest management.

———— 〟 ————

With a long-term perspective, with few loans to pay off, and with no mills to feed, institutional investors such as colleges or pension funds seem to be more appropriate owners of the forests than debt-driven production corporations. Still, they are subject to the impact of interest. Since they must provide their stockholders with competitive returns, they remain under some pressure to practice hasty forestry.

How can the influence of interest be bypassed entirely? There are three possible alternatives: (1) place the forests under the supervision of individual owners who are not eager to trade in timberland for more lucrative investments elsewhere, and for whom profit maximization is not the sole or driving force; (2) control the forests rigidly through regulations which mandate longer rotation cycles, more biological diversity, and larger investments in the future health of the ecosystem; or (3) put the forests in public ownership, to be managed for the benefit of both present and future generations of people, and for the good of the earth. These forms of ownership and control will be explored in the following two chapters.

8

PUBLIC OWNERSHIP
The Workings of Bureaucracy

For better or worse, timber ownership in this country has been partially socialized. The Forest Service controls 18% of the commercial forest-land in the United States, while other federal, state, and local agencies administer an additional 10%. The impact of governmental ownership, however, is stronger than these numbers suggest, since public lands tend to have more standing timber than private lands. The government actually holds 63% of the softwood sawtimber nationwide, most of it within the National Forests.[1]

The impact of the government is particularly strong in the West, where two-thirds of the forests are in public hands. Federal lands, whether forested or not, make up 46% of the land in the western eleven states. While most of the nonproductive regions are administered by the Bureau of Land Management, many of the best timber-producing sites in the world are controlled by the Forest Service. National Forests account for 39% of the whole state of Idaho, 25% of Oregon, and 21% of both California and Washington.[2]

In a nation that embraces the principles of private enterprise, such a large commitment to public ownership is an anomaly. Theoretically, this opens up exciting possibilities. While the timber industry is not structurally suited to pursue balanced and farsighted forestry practices (nor even to maximize production), the government might be expected to do better. Acting in behalf of the public, governmental agencies should be immune to the negative impact of the guiding rate of interest, free of the blinders of the profit motive which operate in the private sector.

Initially, the Forest Service was established to provide insurance against the depletion of the nation's timber resources.[3] While private and corporate owners were exhausting their own supplies, the government would hold on to significant quantities of timber that could be harvested at some point in the indefinite future. Today, public ownership is seen in more sophisticated terms. The contemporary principles of public management are set down in the Multiple Use–Sustained Yield Act of 1960 (MUSY): the forests are to be managed for a variety of uses, including "outdoor recreation, range, timber, watershed, and wildlife and fish." The Forest Service is supposed to make "the most judicious use of the land for some or all of these resources, establishing the combination [of uses] that will best meet the needs of the American people . . . and not necessarily the combination of uses that will give the greatest dollar return or the greatest unit output." The renewable resources of the forest can be harvested, but only at a rate that can be maintained "in perpetuity." And the harvesting activities must be conducted "without impairment of the productivity of the land."[4]

It would seem that the government is better suited than private enterprise to meet this set of goals. A timber company has no reason to manage its land as wildlife habitat; a timber company would be foolish to ignore "the greatest dollar return" as a management criterion; a timber company might have an interest in one or two crops of trees, but "in perpetuity" has no economic translation. The general public, on the other hand, does have an interest in nontimber aspects of the forest, even if that interest cannot be financially expressed. Insofar as that amorphous body called "the public" cares about its own destiny, the government has both the right and the responsibility to manage its land in such a way that our distant offspring can command a resource base undepleted by selfish exploitation for short-term profits.

The goals of the Forest Service were further refined in 1976 by the passage of the National Forest Management Act (NFMA). In the words of one of its writers, the NFMA "sought to put body, sinew, and muscle on the vague language of the 1960 Multiple Use and Sustained Yield Act."[5] Since the environmental movement had achieved considerable political strength by the mid-1970s, the NFMA placed Forest Service activities under stricter constraints. Streams had to be protected from changes of water temperature and deposits of sediment in order that fish habitat could be preserved. The "diversity of plant and animal communities" had to be protected. All logging sites had to be restocked within five years. Harvesting levels had to be consistent with a strict "nondeclining even flow" standard (NDEF), which meant that current yields would have to be sustainable even after there was no more old-growth timber to cut.[6]

The stated goals of the National Forests, as set forth in MUSY and NFMA, are certainly commendable. But are they realistic? Do the everyday mechanisms of the Forest Service insure the practice of well-balanced forestry?

A federal agency charged with the management of 191 million acres of land, the Forest Service functions in an unavoidably bureaucratic manner. Decisions that affect the vast stands of timber in the Pacific Northwest are made by administrators and politicians in Washington, D.C., who cannot possibly have intimate familiarity with most of the forests they govern. Even if these distant managers genuinely care about the fate of the forests, the translation of sound silvicultural principles into viable bureaucratic codes can be a Herculean task. It is difficult to insure good forestry by administrative or legislative fiat.

Consider the question of the rotation cycle: how often should trees be harvested? According to the NFMA, timber should be harvested at productive maturity, which is defined as the culmination of the mean annual increment (CMAI).[7] Before that point in time, the trees are still growing more vigorously than average; afterward, a new crop would grow better than the old. Harvesting at CMAI ensures that the land is producing timber to its maximum capacity.

It sounds good on paper, but in practice there can be a great deal of variability in the application of the CMAI standard. If the CMAI for ponderosa pine is calculated using the total cubic feet of wood fiber as the measure, the rotation cycle is around 40 years; if the CMAI is calculated according to the sawtimber that can be obtained from logs over 7 inches in diameter, the rotation cycle is 90 years; if the CMAI is calculated according to the sawtimber available from 12-inch logs or greater, the rotation cycle is well over 100 years. The higher the quality of wood that the trees are expected to produce, the longer it takes to get there. Indeed, if ponderosa pine is grown for lumber clear of knots and sapwood, the mean annual increment does not reach its peak until the trees are about 200 years old.[8]

Which of these standards should be used? The NFMA is deliberately noncommittal, stating that the CMAI is to be "calculated on the basis of cubic measurement or other methods of calculation at the discretion of the Forest Service."[9] The NFMA, in order to allow reasonable flexibility, also states that timber stands must "generally" reach CMAI before being cut. The Forest Service has interpreted "generally" to mean 95%. This seems fair enough, but because the MAI curve flattens out near its peak, a 5% tolerance can mean a 25-year difference in the length of rotation.[10]

In actual practice, rotation cycles are strongly influenced by criteria that have little to do with silviculture. Because the Forest Service is used as an instrument of public policy, the application of strict silvicultural standards is often modified. National harvesting targets are determined not only by what the forests can produce but by the state of the economy, the rate of unemployment, the demands of household construction, the optimal number of "housing starts," and other political and economic variables.

Government Forestry: From the Top Down

To understand the nature of political influence, we must examine how the *allowable sale quantity* (ASQ) for each administrative unit is determined. In order to harvest timber, the Forest Service must first offer it for sale. Since this requires the expenditure of money, a schedule of timber sales must be included within the budgetary process. But the budget must travel through the hands of many administrators and politicians: the Washington office of the Forest Service, the Department of Agriculture, the Office of Management and Budget (OMB) in the White House, the Senate Agricultural Committee, the House Interior Committee, the Senate Appropriations Committee, the House Appropriations Committee, the full body of the Congress, and the President.

Each of these centers of power has its own reasons for maximizing the ASQ. The Forest Service bureaucracy, as we shall see, has built-in incentives to push for high allocations. The OMB, in the interests of generating revenue, naturally pushes for top cutting levels, at least in the most productive forests. Since the price and availability of timber are key factors in the crucial housing industry, the White House often prefers higher harvesting levels in order to stimulate the economy. Senators and Representatives, always subject to pressure from special interests, like to see the trees harvested so the mills in their home districts will not suffer from a lack of available timber. Indeed, since 25% of the receipts from all timber sales go back into the hands of local governments, the elected representatives from timber-producing districts can ill afford to vote for minimum cutting levels that would deprive schools, fire departments, and other local services of much-needed revenue. (These funds are in lieu of the property taxes which the federal government does not have to pay. The Bureau of Land Management returns 50% of timber receipts to local governments.) And since members of Congress from timber districts tend to get appointed to the appropriate committees, they have a powerful voice in determining the ASQs. During the late 1980s, Congress repeatedly gave the Forest Service even more money than had been requested to put up timber for sale in the Pacific Northwest.[11]

Technically, the allowable sale quantity is only a ceiling, the upper limit of timber sales that will be funded. In practice, the ASQ functions more like a target, or even a quota. In order to further their careers, Forest Service managers operate under pressure "to get the cut out," as they say in the trade. After being told how much they are expected to harvest, they must look to the woods to rationalize the ASQ in silvicultural terms. Their job is a balancing act: to meet their targets while staying within the limits of the various laws that are supposed to regulate their activities, including the NDEF and CMAI criteria, which are specifically designed to limit the levels of current harvesting and to insure sustainability.

In order to perform this difficult feat, foresters have utilized a concept called the *earned harvest effect* (EHE), sometimes called the *allowable cut effect* (ACE). Since future growth rates can presumably be increased by the application of intensive management techniques (improved genetic strains, brush control, fertilization, and thinning), the subsequent harvests will yield greater volumes. Since more will be cut in the future, more can be cut right now without sacrificing the NDEF standard for sustainability. The only problem with this reasoning, of course, is that higher growth rates in the future are in no way guaranteed. How can we be sure that funds will always be allocated to apply intensive management techniques? And even if funding continues, will intensive management lead inevitably to higher yields in perpetuity? Could the increased instability of a simplified ecosystem instead lead to *lower* levels of productivity? At least so far, the increased yields are mostly on paper, not in the forests. In 1978, the Government Accounting Office reported that forest planners had overestimated the EHE in the Gifford Pinchot National Forest by 100% and in the Deschutes National Forest by 250%.[12]

Increasingly, Forest Service personnel are finding that there is no possible way to reconcile the ASQs with other legal mandates to protect the environment and insure sustainability. In 1990 Fred Trevey, the Supervisor for the Clearwater National Forest, wrote to Regional Forester John Mumma: "From these site-specific analyses, we currently project that the Clearwater's resource capability is about 1200 MMBF for the decade in comparison to the Forest Plan's ASQ of 1730 MMBF."[13] Mumma, an administrator in charge of Region One (the Northern Rockies), reluctantly accepted the input of Trevey and other personnel who worked closer to the ground: "My [forest] supervisors and district rangers in the Northern region recognize that we cannot meet my timber targets. . . . I have done everything I can to meet all of my

Titan tongs: meeting the timber target

targets. I have failed to reach the quotas only because to do so would have required me to violate federal law."[14] According to the judgment of Forest Service professionals, the land in Idaho and Montana was simply not able to produce the amount of timber assigned by the ASQs.

Upset that the ASQs were not being met, Idaho's Senator Larry Craig wrote directly to Dale Robertson, Chief of the Forest Service at the time:

> Dale, I am very disappointed with the Forest Service's accomplishment and accountability for timber outputs in Idaho and the Nation as a whole. You have serious management problems that must be addressed. It is my hope you will move to assure targets are met and line-officers are held accountable for targets.
>
> As the ranking member on the Senate Subcommittee on Conservation and Forestry and a member of the Subcommittee on Public Lands, National Parks and Forests, I intend to take increased oversight of the operation of the Forest Service. It is my intent to keep a close watch on Forest Service management and to raise concerns on a regular basis. . . .
>
> The time has come to get down a month by month expectation and accomplishment for each Forest in Idaho and a summary of Region 1 timber accomplishments from the present to the end of the year. I wish to remain informed, on a monthly basis, of the progress that is made. Please forward such a summary to me by the first Friday of each month. Please include in the report what specific action has been taken to reach the monthly target accomplishments.[15]

The "management problems" in Region 1 were soon addressed by holding Mumma "accountable" for not meeting his targets: Mumma was removed from his post. George Leonard, the Associate Chief of the Forest Service, meanwhile assured the timber industry that the Forest Service intended to honor its "commitment to offering the full allowable sale quantity."[16] Clearly, the ASQ was functioning as a quota, not just a ceiling—and the consequences for failing to meet this quota were severe. In the words of forest economist Randal O'Toole, the ASQs "most clearly resemble the [old] Soviet system of management where a central committee determines production targets and local managers are required to meet these targets at any cost."[17]

When the forests are owned by the public, the practice of forestry is inherently politicized. Was Senator Craig's letter a case of "political interference," or was it instead a fulfillment of his obligation to represent the people of Idaho who were in danger of losing their jobs? Obviously, it can be seen either way, depending on one's own political perspective. But no matter how you look at it, the practice of silviculture is not divorced from the political arena. As long as the Forest Service is a public agency, scientists and professionals are liable to be overruled by the people's representatives, who might have only minimal knowledge of how ecosystems work and what the land can be expected to produce.

Time after time, managers who were trained as foresters and scientists are shocked to discover that they must operate in a political world. Tom Kovilicky, Supervisor for the Nez Pierce National Forest, relates a telling incident: "It was the turning point in my education as a public servant when the ransom message was delivered, in person, by an aide from Senator McClure's office. And briefly it said: 'If you want your approved Forest Plan released (so you can begin implementation), just add another increment of trees (20 million board feet) to be available for harvest.'"[18]

Bureaucratic Survival: How to Maximize Your Budget

Back in 1930, enlightened administrators and legislators tried to remove the practice of silviculture from the political arena. Previously, the Forest Service had to beg politicians for reforestation money every time an area was logged. But with the passage of the Knutsen–Vandenberg Act, special reforestation funds were established from the receipts of the harvested timber. The Knutsen–Vandenberg Act is still in effect today. A "K–V" plan is prepared prior to each timber sale, explaining why money is needed and how it will be used. When cash flows in from the sale, the K–V money is put into a special account which can be spent only on the activities stipulated in the plan. In 1976, the NFMA expanded the K–V concept to include other projects within the sale area, such as intensive management to improve future yields or habitat restoration to mitigate the effects of logging.

The idea behind the Knutsen–Vandenberg Act is commendable: to insure that no area can be logged and then forgotten. On the surface, K–V funds seem to increase political independence and local autonomy within the Forest Service. In practice, however, the availability of K–V funding serves as an incentive for Forest Service administrators to increase their timber harvests—and to practice clearcutting, since the higher expense of reforestation will bring in more K–V dollars. Particularly in times of decreasing discretionary funding, statutory funding such as that permitted by K–V becomes more and more appealing. Although Congress is often reluctant to allocate money for nontimber resources, it is more willing to provide money for timber sales. But through K–V funding, this timber money can be redirected toward other uses. If a District Ranger wants to put in a trail, for instance, he can get the money for the project by harvesting timber near the trailhead and including the trail work in the K–V plan. In the Gallatin National Forest, timber was put up for sale in order to get the money to improve grizzly bear habitat. In the Sequoia National Forest, trees were cut down in order to finance a fire hazard reduction through brush disposal funds, which function much like K–V.[19] Money is available for all sorts of worthy projects—if only some timber can be harvested nearby.

Since Forest Service administrators at every level receive a portion of the "overhead," they can all increase their budgets through K–V funding. Typically, the upper levels of bureaucracy—the Washington office, Regional office, and Forest Supervisor's office—receive about one-quarter of the money generated by K–V plans.[20] This money is not distributed, however, until the plans are actually executed. Consequently, pressure is placed on field officers not only to prepare lucrative K–V plans but also to spend all the money that was allocated within the plans. In the words of economist Randal O'Toole, "If trees are not planted in California, a Forest Service official in Washington may not get a desired pay increase. If herbicides are not sprayed in the Klamath National Forest, an administrator in San Francisco may lose his or her job. Silviculturalists are pressured to meet reforestation and similar targets to maintain the flow of cash to higher levels of the bureaucracy."[21]

Ironically, the ASQ quotas are mirrored in these K–V and brush disposal "targets," since in each case bureaucratic structures seem to reward high levels of timber harvesting. But this internally generated budgetary inflation does not always coincide with the best interests of the forest. Often, activities are pursued simply for the sake of spending the appropriated funds. According to a study in the 1980s, 43% of the district silviculturalists stated that some of the herbicide release they had performed in the preceding five years had been either ill-timed or unnecessary. Similarly, 20% stated that some of their site preparation projects had been either ill-timed or unnecessary, while 23% admitted that they had actually falsified site preparation data in order to meet their targets.[22] Unfortunately, budget maximization seems to function as a motive force behind silvicultural strategies.

The unintended side-effects of the Knutsen–Vandenberg Act are hardly unique. It is not uncommon for well-meaning legislation to become counterproductive as it filters through a huge federal bureaucracy. For instance, the Resources Planning Act of 1974 (RPA) requires the Forest Service to prepare a national plan every five years. It sounds good enough; certainly, all activities should be coordinated, and they should be made to conform with basic objectives. But the planning process has liabilities as well as assets. The Forest Service bureaucracy operates on four simultaneous levels: national, regional, forest, and district. According to the RPA and the NFMA, plans have to be prepared not only on the federal level but also for the 9 regions and 156 National Forests. Each of the upper three echelons of the bureaucracy must therefore engage in huge amounts of paperwork before anything at all can happen down on the ground. The preparation and updating of these parallel plans consume an immense amount of institutional energy, inevitably weighting the agency with midlevel bureaucrats rather than people who work in the woods.

Even on the district level, where workers presumably are in touch with the actual land, the need to feed the machine is steadily increasing. According to one District Ranger, "About 50% of district resources (people and time) are spent providing information, computer data, writing or rewriting reports to satisfy the needs and whims of the massive internal Forest Service bureaucracy."[23] And when the budget crunch hits, middle managers on the national and regional levels are *not* the first to go. Field workers, on the other hand, are more expendable. The list of "surplus" employees includes a disproportionate number of low-level, on-site workers. Upper-level managers, meanwhile, fare better; the number of Forest Service employees making over $80,000 actually increased from 83 to 124 during 1992.[24]

Ironically, a career in the Forest Service these days does not generally entail spending much time out-of-doors. Most of the real work—harvesting timber, planting trees, repairing streams—is performed by private, contracted labor. But these private contractors have no personal stake in the health of the forests in which they work. Since they are not government employees, their only responsibilities to the public and the land are the narrowly defined stipulations within their contracts. Literally, they are here today and gone tomorrow, mere transients in the woods. No matter what type of task is being performed, the primary object of the contractor is to meet (or pretend to meet) contractual obligations with minimal effort in order to maximize profit.

Since contractors have no vested interest in the overall health of the forest, they must be made to behave as if they did. Forest Service administrators therefore stipulate an elaborate set of rules that must be followed. Timber sale contracts, for instance, are voluminous documents that prescribe the details for road construction, logging techniques, the removal of debris, and so on. And yet, although the work is severely circumscribed by contractual specifications, the infinite variety of circumstances, combined with the decentralization of most activities, insure that there will always be some room for a contractor to bend the rules. Since the concerns of the workers and contractors do not necessarily coincide with the concerns of the forest managers who make the rules (let alone with the needs of the forest itself), some sort of self-serving finagling is bound to occur.

With the actual people in the woods regarding contractual stipulations as "management constraints" that are to be reluctantly obeyed or surreptitiously avoided, an entire bureaucracy of federal employees is necessary to ensure that the letter of the law is enforced. There are log-scalers to measure the quantity of wood that is taken, field inspectors to make sure the work is performed in accordance with specifications, and office personnel to coordinate and oversee the work of the various on-site inspectors. Each federal inspector is "boss" to the loggers or other contractors, who understandably resent having so many overseers. This fosters a certain alienation among the workers of the woods, and it is not uncommon for this alienation to be expressed by willful disobedience or neglect.

Raymond Cesaletti

Reforestation Worker

"I'd like to go back to the beginning and mention how I got into tree planting. I always had an inclination for things agricultural and for the woods. So the first thing I did when I hit the Northwest was pick fruit: apples, pears, and peaches. Then when that season ended, in the fall of '71, I applied to the Forest Service. You know, even back when I was a Boy Scout, I always wanted to be a forest ranger. I didn't get hired.

"Then I went to the employment office and took a job planting trees. I had no idea what it was going to be. I thought we were going to go out to a farm with a wheelbarrow and little trees in burlap bags. I didn't have any raingear or equipment. I showed up at 6:30 one morning and got into a crummy with a bunch of wet-backs. Couldn't see out the windows because of the mud on the outside. We rode for what seemed like hours. Finally we got out and there we were, on the side of a cliff, practically. It was pouring rain.

"We were all greenhorns, right out of the employment office. So they gave us a quick lesson, a six- or seven-minute training session, and sent us down the hill. I didn't know what the hell I was doing. All the trees I planted were bad. I could barely walk around, let alone chop a hole in the ground and plant a tree. Made $8 that day. It was the worst day of my life. The other two guys I was with quit. They just didn't come back the next day.

"But I started watching those Mexicans. They were making about fifty bucks, so I said, 'There's something going on here, and I'm going to find out what it is.' I got taught some of the tricks by the foreman of the crew. He took me aside and said, 'Look, here's how to hide 'em.' I got a very specific lesson in what they call the 'Wino Technique.'

"The main thing you have to learn is how to fool the inspector. Like if you're in nasty ground that's really not suitable for a bare-root seedling, if it's all rotten wood with solid rock underneath it, you're not going to dig a hole and plant a live, two-year-old tree that's going to survive. But you can get them in there and make them look decent. To somebody who walks by, it'll look like a nice, pretty, little tree standing up straight. Of course in two months it'll be dead. Or you can cut the roots so the tree will fit in a smaller hole.

"I play it by ear. Whatever the circumstances dictate, that's what I do. If a job is very high paying and the standards are very high, I will produce the highest quality work you could possibly imagine, a perfect job with no corners cut, no fudging anywhere. But if the standards are high and the pay is low, too low for those quality standards, then I will fudge where I have to make my money. Let's face it: I have a right to live and to eat. That's when I work for a contractor. If I work for a cooperative, it's different. I'm not going to fudge on working for a cooperative, because I'll only be hurting myself.

"Once I had a job on a contract for Weyerhaeuser. The contractor was getting paid a piece rate, but he was paying us by the hour. So there was no stimulus for the crew to hide trees, and he was wondering why he wasn't making out so well. I was running the crew for the guy, and we wanted higher pay. He said, 'Well, production isn't high enough to pay more money.' I just said, 'Yeah, because we're trying to give you good quality out here. If you want production to go up, quality is going to go down.' But he said if we got the crew average up, then we'd be likely to get more pay. So then I went out and started jacking up the crew average. I grabbed a few of my cohorts and taught them the tricks, and we started padding the averages to try to get the pay up. I taught them how to hide trees the professional way.

"I showed them the slit method, or envelope method. It involves getting in a place where there's low, tight ground cover. Salal will do, but it's not the best. Never where there's bare dirt. You need a tight root system that's in the humus. You stick the hoedad in just below the surface to make a slit parallel to the surface of the ground, about an inch or so down. You pull up a little to open up that slit. You reach into your bag and pull out three to five trees, bend them in half, slide them into the hole, and lightly push the slit closed with your foot.

"If you do it smoothly, it looks just like you're planting another tree. We once had a bet with the foreman, him and the inspector. We were hiding them, they knew it, we knew that they knew, they knew we knew that they knew. I made a bet with them that they wouldn't find a single hidden tree. And they didn't. The two of them were digging like maniacs all morning, and I was laughing my balls off. I must have hid about 150 of them within two hours, me and the guy behind me. They never found one. There was salal and salmonberry flying all over the place. Those guys worked harder than I ever saw them work, just trying to find where we hid the trees.

"The Forest Service mostly pays by the acre, not by the tree. That leads to a special type of cheating. You, as a contractor, are not going to want your crew to plant extra trees. You're not worried about how many trees per acre survive, you're just worried about covering your ground. You want your crew to get the most amount of acres they can in the least amount of time, which means: 'Don't plant any extras.' Of course you can't plant too few, either, or they'll bust you for it. But sometimes a contractor will have a couple of top planters, his hot dogs, who will get up there where the inspectors can't even keep up with them. They'll wipe out a whole bunch of ground real quickly. Nowadays it's more likely that the inspector will get up in there, but it used to be that they didn't.

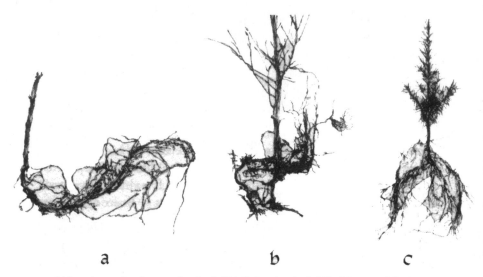

a   **b**   **c**

(a) J-rooting, causing slow growth or death; (b) balled rooting, dead within three years of planting; (c) a good root structure

Those hot shots knew how to go out there to the far end of a unit and chop out a bunch of ground. It would get planted, but in a token fashion, a few here, a few there. Dig it up, make marks and tracks so that it looks like you've been around it all, and then get the hell out. Don't even let the rest of the crew get in there, because they'll go in there and start planting trees.

"Now I'm not condoning this practice, and I don't condone stashing trees. I hate it. It stinks. It feels bad. But if I'm out there seventy miles away from my family for the day, working in the stinking-ass rain and wind and everything else, I'm damn sure gonna get my wage out of it, no matter what I have to do to get it. If the foreman had the money in his pocket, I might be tempted to hit him on the head and take my wage home.

"But I do believe in reforestation. The old-timers, when they used to go into Lucky's downtown and say, 'Who wants to plant trees today?' and haul out the guys who were drinking at eight o'clock in the morning, they didn't give a shit. They just wanted the money. I've got a college education. Why go out into the woods and plant trees? It's stoop labor, nasty weather conditions, and everything else. But I've always felt, and I still feel, that it's a worthwhile thing to do. You feel good when you're done with it, even when you've cheated at it. So maybe you've fudged 10% of the day. You cheated. But the rest of your day, 90% of that day, was a noble effort to replant the forest. I mean I can go now and show those places to my kids. The trees are six feet high. I mean, to me that's meaningful. Most of the men feel like they're providing something that's needed—and they are. It's almost an extension of the Boy Scout mentality, which gets a little disgusting sometimes, but that's the way it is.

"But I won't knock the guys who are just into cheating for quick production all the time. Sometimes it's almost impossible to produce what they want, yet you still have to make your money. You are not out there purely as a charitable gesture. You're trying to make a living at it, see? So when they're asking you to do the impossible, which they often do, what are you gonna do? You do what it takes to make it look like you're performing the impossible.

"The most common way of cheating—and it's done all the time—is J-rooting. Simply speaking, a J root occurs when you open up a hole and shove the tree down in it, so that the roots form a 'J' against the bottom of the hole. There are many variations: true J-roots, curled roots, bunched roots. Anything other than the root hanging in a natural attitude with the dirt fluffed in around it can be called a J-root. If you look at any Boy Scout manual, they show you how to plant a tree. They show 'right' and they show 'wrong.' And a J-root is 'wrong.'

"The Forest Service now, because of all the hassles they've had with contractors in the past, has several pages of instructions to inspectors, defining how the inspector must go about digging the tree to find the J-root. It's so long and involved. A private-industry inspector has a lot more autonomy. A government inspector has a manual that he has to follow, and if he doesn't do things exactly by the book, the contractor can call him on it and file a claim with the government. The government specifically describes how many plots the inspector has to take per acre, how many trees in each plot he has to dig.

"And they'll tell him exactly what is involved in digging a tree. He'll get downhill from a plant, go out about a foot, and start excavating. He digs down a good two feet, and then he takes a sharp stick or something and starts very carefully undermining the soil to dig in under the roots of the plant. Hopefully, he'll uncover the roots to the point where he's looking at a cutaway of the planting. But quite often he'll blow it. He digs and digs and his stick catches one of the little roots, and he pulls the whole tree out.

"The Forest Service technique, if they always did it properly, would prevent the inspectors from keeping up with the crew. It's difficult to expose all the roots of a new plant, which has a delicate root system anyway, without disturbing the way those roots are sitting in the ground. What happens is that the inspectors often fudge. If an inspector doesn't find anything wrong, they're going to lean on him. They know that people cheat. So he just fudges it. He calls most of them good, but he says, 'Well, I found a bad one over here.' The inspector can be lazy just like anybody else.

———— ” ————

Bureaucratic Survival: How to Get Ahead

The government inspectors, like the contractors and workers, have no vested interest in tomorrow's timber. Like the people whom they are employed to judge, the inspectors are ordinary working folk who might have more of a stake in their paychecks than in the actual survival of the seedlings. And the inspectors themselves are relatively powerless, performing their jobs according to sets of specifications that have been handed down from above. They too are subject to continuous scrutiny. Because of the hierarchical structure of the Forest Service, personnel within each stratum must pass judgment on the workers beneath them. With the exception of those on the lowest rung of the ladder, everyone is both a boss and an underling.

How does the bureaucratic organization of the Forest Service affect the work of its own employees? Theoretically, Forest Service staff, unlike private contractors, are responsible to their employers: the public. In practice, however, they are responsible first and foremost to their immediate superiors, who subject them to frequent evaluations. Since promotions can only be obtained by pleasing one's superiors, the personal ambition of each worker easily becomes wedded to the policies of the organization. According to Glen Robinson's in-depth study of the Forest Service bureaucracy:

> While the Forest Service can and does tolerate a variety of views on particular issues and on particular subjects, it does attempt through its hiring, assignments, and promotions to develop loyalties to traditional policies of land use and management. While the agency does not consciously attempt to discourage innovation or new ideas, the incentives created by

its emphasis on internal promotion and loyalties to institutional values favor a fairly conservative and stable policy of land use management.[25]

Institutional loyalty is also achieved by a policy of frequent transfers. By continually moving its personnel from place to place, the Forest Service ensures that a worker's primary allegiance is to the organization, not to a specific locality. A worker who chooses not to transfer from his place of work is unlikely to be promoted; consequently, few lay down permanent roots. According to personal accounts, the average stay in a given location is five to seven years. A few years back, a telling memo was circulated by a Regional Director of a similar federal bureaucracy, the National Park Service: "It has come to our attention that Park personnel are refusing transfers because they want to be happy. Happiness is not always in the best interest of the National Park Service."[26]

The consequences of this transience are profound. The managers of our public forests have no direct, personal stake in the long-term productivity of the land that is placed in their care. Just as the contractor may have more of an interest in personal profit than in the long-term results of the work he performs, so may the forest manager have a greater stake in the advancement of his professional career than in the fate of our future forests. Certainly, there are many forest rangers who seriously care about the quality of their work; but the bureaucratic structure does little to promote a lasting concern for the health of specific localities.

The conservative structure of the Forest Service tends to perpetuate traditional policies and priorities; the people who advance within the ranks are those who identify most closely with these policies and priorities. And historically, the traditional approach to American forestry has emphasized resource extraction. A 1981 survey of 400 high-level Forest Service managers found that the responses closely resembled those of industry representatives (timber, mining, and grazing), while differing significantly from those of environmentalists.[27] An earlier study found that 71% of Forest Service professionals believed that utilizing resources was their most important mission, while only 17% thought that the preservation of resources or the maintenance of natural beauty should take precedence over active resource utilization.[28]

And the primary resource that these foresters wished to utilize was certainly timber. According to the Multiple Use–Sustained Yield Act of 1960, the extraction of timber was to be only one of several equally important uses of the National Forests. In practice, however, timber has continued to be "more equal" than the others—despite the dictates of "multiple use." A subsequent management plan for the Willamette National Forest, for instance, appeared on the surface to take MUSY rather seriously: "Let it be understood, therefore, that all resources on this forest are of equal importance in the basic concept of multiple-use management, and that no one resource shall be allowed to assume an over-riding position." But the plan went on to state that good timber management was the "most useful tool for enhancing other uses." Logging roads, the plan stated, provided access to hunting and berry picking in the "openings" of the forest. Logging of old-growth was supposedly good for the soil, since it "can reduce soil impacts from uproots." Logging was even supposed to enhance the esthetic value of the landscape: "Clearcuts break the monotony of the scene, and deciduous brush in these areas furnish fall color and spring flowers for at least 10–15 years."[29]

During the 1980s, the Forest Service's emphasis on timber was challenged repeatedly by outside interests. Using the judicial system as a battlefield, environmental organizations filed suits which claimed that Forest Service policies violated the National Environmental Policy Act (NEPA) and the Endangered Species Act, as well as the multiple-use provisions of MUSY and the biological diversity provisions of NFMA. The legal assault on timber domination within the Forest Service came to a head in 1989 when Federal District Judge William Dwyer granted a preliminary injunction on 135 timber sales that were said to threaten the habitat of the Northern Spotted Owl. According to the court ruling, which was upheld on appeal, logging could not continue until the Forest Service revised its spotted owl guidelines to provide for more protection. Together with BLM, the Fish and Wildlife Service, and the National Parks, the Forest Service then convened an Interagency Scientific Committee to develop a comprehensive spotted owl strategy. Headed by Jack Ward Thomas, the wildlife biologist who would later become Chief of the Forest Service, the committee called for large tracts of old growth to be preserved as "habitat conservation areas."

The first response of the Forest Service was to develop a new plan "not inconsistent" with the Thomas report. Because of a cabinet-level directive from the Bush Administration, however, a scientifically credible plan never materialized. In 1991, Judge Dwyer directly prohibited any further sales in spotted owl habitat areas. Dwyer's second ruling was particularly forceful, scolding the Forest Service for its "deliberate and systematic refusal" to comply with federal laws. Dwyer, a Reagan appointee, also implicated the current Republican administration, placing the blame for

Poised to pounce: the spotted owl

Forest Service negligence on "decisions made by higher authorities in the executive branch."[30] [The social and economic implications of the Dwyer rulings will be discussed in Chapter 10.]

Although the spotted owl controversy dominated the headlines, there were other significant challenges to the timber policy of the Forest Service in the 1980s. Not only did environmentalists object to logging the lush and productive old-growth forests in the Pacific Northwest; they also objected to logging less productive stands in other areas. In the Rocky Mountains, for instance, trees often grow on fragile and sensitive soils that cannot easily withstand the impact of logging. Since timber grown on these marginal sites is generally less productive, the income generated from harvesting these stands is often less than the expenses which the Forest Service incurs in putting the timber on the market. These *below-cost sales* (BCS) have been heatedly contested by environmentalists-turned-economists who cannot see the wisdom of logging lands which are so marginal they can't even turn a profit. With the exception of Region 8 (the Southeast), Region 6 (Oregon and Washington), and a portion of Region 5 (northern California), timber sales from the National Forests during the 1980s were notorious money-losers, costing the taxpayers about $150 million a year.[31] The idea of jeopardizing the environment in order to lose money didn't seem to make any sense.

In defense of the sales, the Forest Service argued that it should not be required to use the same standards of profitability that are applied to the private sector, since many of its costs were mandated by federal law. In order to meet multiple-use objectives, it had to prepare extensive plans dealing with such variables as wildlife habitat and aesthetic features. The cost of protecting the environment was basically being charged to timber, and this is what made the sales lose money. It did not seem fair, by this line of reasoning, to curtail timber sales because of the costs of meeting nontimber objectives. Environmentalists, on the other hand, argued that the costs of maintaining wildlife habitat or aesthetic features *should* be born by timber, since the need to protect these resources would not exist if they were not threatened by logging. Below-cost sales, by their line of reasoning, represented a true loss: timber could not pay the price of protection.

Timber interests, with some justification, wondered why environmentalists applied the standard of profitability to logging but not to other uses of the forest, such as camping and hiking. If timber had to pay its own way, why should yuppies get to wander through the woods for free? This argument elicited two distinct responses. Some opponents of BCS agreed with the analogy, admitting that *all* National Forest consumers, including recreational users, should bear the appropriate costs. Others, however, made a distinction between environmentally destructive uses, such as logging, and environmentally benign uses, such as hiking. Timber harvests, they maintained, should make at least enough money to pay for the damage they caused.

The debate over below-cost sales was ironic: environmentalists usually hesitant to use economic reasoning suddenly showed a serious concern for the well-being of the federal treasury, while the timber industry, strangely enough, pointed repeatedly to the section in MUSY stating that decisions were *not* to be made using "the greatest dollar return" as a criterion. The two sides had basically flipped sides in order to support preconceived positions. The real issue, of course, was not about the national economy but about harvesting timber in environmentally sensitive areas that were simultaneously more fragile and less profitable than commercially lucrative sites. As one industry representative observed, environmentalists were "taking advantage of a far more fiscally conservative public climate in proposing a new tool to constrain commodity management on public lands."[32]

For many years, the Forest Service sided with the timber industry on the issue of below-cost sales. Even where timber could be grown only marginally, it continued to dominate other uses; trees were cut on lands which might have been more appropriately allocated to wildlife, fisheries, or recreation. The resistance put up by the Forest Service to criticism of its BCS policy was an indication of its continuing emphasis on harvesting timber. In 1993, however, the Forest Service suddenly changed its mind, proposing that BCSs be phased out on one-third of the National Forests. In part, this turnabout was due to a new administration in Washington; President Clinton had vowed to make the federal land agencies more environmentally responsible—and more financially sound. But the new policy also reflected deep-seated changes within the Forest Service itself.

A Challenge: The Times They Are A-Changin'

Despite the conservative, timber-dominated orientation of the Forest Service bureaucracy, an increasing number of employees in recent years have begun to question the traditional values of their agency. In a survey of top "line officers" and new recruits, respondents indicated that the values that *should* be rewarded by the agency were: (1) professional competence, (2) care for healthy ecosystems, (3) care for future generations, and (4) innovation. These same respondents then listed the values that *in fact* most rewarded: (1) loyalty to the agency (2) meeting targets (such as ASQs and K–V spending), (3) promoting a good Forest Service image, and (4) following rules and regulations.[33] Clearly, there was a large discrepancy between the personal values of the Forest Service staff and the actual workings of the machine. The stage was set for political conflict inside the agency.

In 1989 Jeff DeBonis, a timber sale planner from the Willamette National Forest, decided to speak out against what he called the "cavalier attitude" which prevailed at the time:

> I became very disgusted with the amount of damage that I was seeing in conjunction with the logging: riparian destruction, erosion, landslides. I would go out and look at the

ground and make recommendations and write entire reports and turn them in to my superiors, and they would say: "You didn't cut enough timber." They would want more timber to cut than I could find. I was having a crisis of conscience, so finally I decided to go public.[34]

When DeBonis aired his complaints, he received support from numerous co-workers in other National Forests. In order to provide an appropriate voice for dissent within the agency, he founded a group called the Association of Forest Service Employees for Environmental Ethics. The AFSEEE vision was to "forge a socially responsible value system for the Forest Service, based on a land ethic which ensures ecologically and economically sustainable resource management." By 1993, AFSEEE had expanded to include over 2,000 Forest Service employees among its 10,000 members, or 5% of the total agency staff. Unfortunately, many of these workers still felt their dissent could place their careers in jeopardy.

Rich Fairbanks

Timber Planner, Willamette National Forest

"About 1970, I got hired onto a fire crew down around Susanville. I really liked it, so I got one of those two-year technician degrees, an associate of science in forestry. I moved up to the Olympic Peninsula, burning clearcuts. We believed we had to burn all of our clearcuts because there was so much waste wood left in there. It was a crude imitation of natural succession, but it did the trick.

"That was my first real exposure to hardcore, industrial forestry. I worked in the Shelton Ranger District, where they were cutting 140 million board feet a year off a 120-thousand-acre district. Just mowing it down. I quickly realized, as most of us did, that they were cutting way too fast. You really didn't need a FORPLAN run or somebody with a Ph.D. to validate the fact that they were going to run out way before anything grows back. And they did.

"They did a lot of side-cast road construction on real steep country, eighty percent slope. (Sometimes we'd have to hang a rope to be able to work on it.) They'd scoop out the part where the road was going to be, then instead of hauling it back somewhere to use as fill, they'd just dump it over the side. At one time, there were a lot of anadromous streams; they used to have some nice fish in there. Well, the side-casts never stabilized, and all that exposed sub-soil rattling down the slope just wiped out the fish.

"I was a crew boss for the burning, then in the off season we'd plant trees, do thinning, whatever. When it got too dry to burn, they'd send us over to eastern Oregon to fight brush fires. For awhile I was running the slash burns, deciding how to burn and when to burn. Then about 1980 I went over to hotshot crews, supposedly an elite in fire fighting. I did that for four years in eastern Washington and over to Mt. Hood in Oregon. That was a lot of fun. I got to travel around, a lot of cheap thrills—being let off in weird places, then they come in and pick you up a few days later.

"But in the Forest Service, if you don't have a four-year degree, you're pretty disposable. So I went back to school at night during the winter. I got a bachelor's degree, and finally in 1984 they offered me a chance to do some work on a forest plan. It turned out I was good at it, so I continued to do that for five more years. Under the National Forest Management Act, we were supposed to do these scientifically and politically credible management plans. We had to map out what parts of the forest we would put to what use—wilderness, full cutting, riparian, whatever it might be. There's also a temporal component, harvest scheduling; over time, we're going to get x amount of commodity out of this particular land allocation.

"That's when I started to realize how much political interference there was, pressure to exaggerate the amount of timber we could get. I worked as a writer-editor, so I was privy to a lot of documents and saw a lot of what was going on. I'd sit in meetings and take notes. It was real easy to figure out that the good stuff in your career, the real perks, came when you got more boards. You pushed the envelope, and it was always in that one direction: a little more timber. The people who were getting promoted were the ones who went along with this management scheme.

"I believe that pressure existed throughout the Forest Service, and it created an artificially high level of timber harvest at a time when the science and the empirical knowledge of the people in the field were pointing in the other direction. The people who worked in the woods were saying, 'Hey, these creeks are getting muddy. Hey, we're having problems with bugs. Our fires are getting crazier because of the fuel build-up.' On timber sales, you'd have a good group of people getting together from different disciplines, scientifically and technically trained folks who know what they're doing.

They'd do a lot of walking around and finally come up with a fairly complex analysis of the conditions on that particular piece of ground. They'd put together a set of alternatives, and then the District Ranger would pick one of those alternatives. He'd almost always pick one that had a lot of volume in it. And then he'd sign a paper saying, 'There is no significant impact in this sale.' They'd sign these papers time after time, and then, after twenty years of saying 'no significant impact,' the spotted owl is suddenly in trouble. There was an attitude of: 'We will get the cut out, and anything that gets in the way of getting the cut out is not good.' Anytime someone out there in the woods would identify something that was a problem, *he* would become the problem; he would be looked at as disloyal, a troublemaker.

"I had a ranger tell me, 'You will do an EA (that's an Environmental Assessment for a timber sale) every four months.' He wanted me to crank these suckers out. Well, nobody's ever called me lazy (I used to get cash awards for working real hard), but I did want to walk enough of the ground to know if there's any surprises out there. By this time I had become a timber sale planner, and a big part of that job is to know what's feasible to cut. A lot of it is steep ground and there's creeks. It's big old growth, so it's heavy timber. The way you drag it up to the landing could be damaging. It seemed obvious to me that every piece of ground was different, so you really had to plan it out pretty well. But the attitude was just: 'Do a sale every four months. Get the cut out.'

"Working on forest plans, I started to realize that we really were pretty far from what the public wanted us to do—and also from the law. I had been out in the woods all these years saying, 'Damn, they cut too much.' But so what? I thought that's just the way things were: the MFWIC's [mother_____ what's in charge] make stupid decisions, and then you have to implement them. Everybody I worked with said, 'Of course they're screwing up. What d'ya expect?'

"But as I got more education (by this time I had even had some graduate work) and I got higher up and saw how the plans were formed, I started to think, 'This is scandalous.' When you get to be white collar, you start having illusions that somebody gives a shit what you think. I thought, 'Maybe I can actually do something about it.' You'd be surprised how many people who are into forest planning at that level, who look very bland and mouth whatever they're supposed to mouth, are just furious inside about the way the forest is being abused.

"By '89, shortly after moving down to the Willamette National Forest, I realized that I was going to have to do something. You really have to choose up sides. This is the kind of area where you can't stay neutral very easily; it's very polarized socially and politically. And the Forest Service is very arrogant: 'You gotta be on our side, or you're the enemy.' When they force you to make a choice like that, you go ahead and choose.

"Obviously a lot of other people were feeling that way, because AFSEEE was being formed right then. There's good reasons that AFSEEE was born on the Willamette. This was an exceptionally arrogant management team, and it was also the highest cut in the nation, 700 million board feet a year on this Forest—in fact, one year they cut over a billion. So they were really whacking. 'Whack, stack, black'—clearcut, yard it, burn it. They call it the 'flagship' of the Forest Service, because it has the highest cut. They were very, very focused on getting the volume out—so focused that they ignored a lot of things that they should have been paying attention to. So now, of course, the flagship has run aground and they're downsizing, throwing deckhands overboard.

"I signed up with AFSEEE within about six months of its founding, very quietly at first. We were getting a lot of interested people, but nobody over about GS 11. Somewhere around $30,000 a year was where the cutoff occurred; above that, you just didn't see people who were angry enough or willing to risk that nice salary. It was mostly field people; when you're in the field a lot, you really tend to see the result of policy. If you're gonna get pissed, it's when you actually see the mud in the creek.

"So we started going around to timber sales, a group of us, and we'd write it up. We could write it up better than anybody. We'd have an engineer go look at the road, somebody who's been looking at roads for ten years. We'd have a specialist in wildlife biology go look at the elk herd. (All on our own time, of course. We didn't do anything with government equipment; when you joined this group, you quit taking pencils home.) Then we'd start writing letters: 'Here's exactly what's wrong with your sale.' Real technical stuff. And we of course shared those with environmental groups. I guess it was pretty effective, because some of those sales didn't go forward. Not that we didn't believe in timber harvesting; we just thought it should be done intelligently.

"Many of us didn't sign our names; it was just too risky. I don't think there's any organized conspiracy against AFSEEE, but it could still be unpleasant. Some people who came out of the closet had no retaliation of any kind, but there's also people who had a lot of retaliation. It depends on the personality of the immediate supervisor and maybe the person over them, or even how ornery your own personality is.

"When I finally started coming out of the closet a little bit, things started to happen at work. I gave an interview to CBS News on my own time, and I made it very clear to CBS that I was talking from my own opinion. (Unfortunately, the editors failed to make this clear.) I said some very mild things about log exports—certainly a lot milder than what I'm telling you tonight. Two weeks after it aired, I was declared a 'surplus employee.' After twenty years with the Forest Service! I was on a fully funded project, so I knew there was money. But they said, 'Sorry, you'd better take another job.' I moved over to timber theft, but I filed a grievance and the union's been real good. (All those years of paying six bucks a shot really paid off.) I got temporarily reassigned to my old job but at a lower level—from an 11 down to a 9. At least I didn't have to move, and I've got a boss who's just great. So I'm back where I want to be, at least temporarily. But it's up in the air; it hasn't been resolved yet.

"Obviously, being involved with AFSEEE is not a good career move. But that's okay. I'm a GS 9, I make my mortgage every month, I'm doing what needs to be done. Some people see it as disloyalty, but I see it as necessary dissent. It's unfortunate that some people think it's bad.

———— ” ————

The move for reform within the Forest Service, at least to some extent, is reflected by changes in national policy. In 1990, the Washington office announced a "New Perspectives" program intended to redefine land management strategies. New Perspectives gave official recognition to the ideas of land stewardship and ecological diversity; it also promised to bridge the gap between the scientific community and on-site land managers. Perhaps most significantly, it seemed to invite public participation in the actual workings of the Forest Service, encouraging partnerships between environmentalists, timber workers, and the agency staff in developing programs that would simultaneously provide jobs and treat the forests respectfully. This vanguard program was followed in 1992 by the landmark proclamation, issued by Forest Service Chief Dale Robertson, announcing that "ecosystem management" would henceforth serve as the guiding force of *all* agency actions. At least on paper, the era of timber domination appeared to be coming to a close, to be replaced by an "ecological approach" to multiple-use management which would maintain "diverse, healthy, productive, and sustainable ecosystems."[35]

Is the redirection of the Forest Service for real, or is it simply a public relations maneuver to court the support of an increasingly hostile public? Some of the New Perspectives showcase projects seem to have fallen short of their goals. In the Shasta Costa basin of the Siskiyou National Forest, local forest activists were invited to help prepare a three-year program that provided for environmentally sensitive logging in a largely roadless watershed. The partnership between activists and the Forest Service broke down, however, when it became evident that the slight reduction in harvesting would have to be compensated by increased cutting levels after the project ended in order to achieve the "timber outputs" called for in the ten-year forest plan. The showcase, in the minds of some of its participants, had regressed to business-as-usual.[36] Any hope of salvaging the partnership disappeared when crucial Forest Service personnel were transferred out of the district.

Elsewhere, however, there is much hope for a new direction. In Oregon's Mt. Hood National Forest, the Clackamas Ranger District is experimenting with a stewardship program and serious community involvement, hoping to put the Forest Service back in touch with both the land and the people.

John Berry

District Ranger

"We're trying to move towards ecosystem management that includes humans in the environment. In order to do this, we're looking at different approaches to our organization and institutional culture. On every Forest Service assignment that I've ever had, from Wisconsin to Idaho to Oregon, in the coffee room there's always been informal discussions about the concept of 'stewards.' For my entire career, it's been in the back of my mind. Yet because of our functional budget and internal politics, it's not very feasible and rarely tried.

"A few years ago, some of us took a trip to Collins Pine down in California. What struck me the most was that an area about a third the size of my district was managed by just four employees. At that time we had 76 full-time employees on 200,000 acres, and in addition to that, we had another 70 seasonal employees—and here they were managing 80,000 acres with only four people.

"Of course, we have to do some things that private owners don't have to do: NEPA planning and NFMA requirements, for instance. We have to have specialists preparing reports and analysis. But there's other things that we just do to ourselves, like inventory work. We asked them, 'How do you figure your sustainable yield?' They said, 'Well, once every five years we go to our permanent fixed plots and measure them. It takes about a month of one person doing that. We calculate what the standing volume is, what the difference in growth was, figure in what we harvested off, and that sets the harvest for the next five years. The calculations take about four hours.' Contrast that with our own FORPLAN process, where it took us 12 years and somewhere around a million-and-a-half dollars, plus all the satellite inventories and so on. Those are things that the National Forest Management Act doesn't require, but that's just the way the Forest Service internally does its business, from Washington on down.

"Even with such a small staff, Collins Pine was doing an excellent job of management. Their land tastes, feels, and looks like a forest. They were doing it largely by the paint gun [selective cutting, with trees individually marked]. They take out a fairly high volume, yet they still have more standing volume on their industrial land today than when they started 50 years ago. They're definitely on a sustainable program.

"I came back and said, 'Why can't we do that?' A lot of us in the Forest Service have wanted to move away from clearcutting and towards a stewardship idea, but what keeps us from doing that? It struck me that there were some organizational barriers to it. We're organized around an assembly-line fashion, where one specialist deals with a project and it moves along and passes to another specialist, and then it passes on to the next and the next and the next. No one person ever sees the whole picture. We have people who are really excellent at reforestation; they would match up against anyone in the world. They know how to grow the best, superior stock in the nursery, they know how to handle it and get it planted out there so it will survive. But they have no idea how that fits in with the surrounding projects, like the fish and riparian projects and everything else that's going on around them. All the projects are interrelated—we intuitively know that—but the specialists aren't responsible or accountable for the whole system. Instead, we manage by covet and conflict, with the specialists competing against each other for resource dominance: 'What's my resource is mine, and what's yours is up for negotiation.'

"Finally, we decided to institute a stewardship program here in our district. The idea was to break some employees out of the specialist mode, and to have them focus on a small enough area where they aren't trying to look at all 200,000 acres. We moved them down to a specific drainage where they had at least a reasonable chance of getting to know that piece of ground fairly intimately. The stewards were responsible for keeping track of all operations, for maintaining a central pool of all resource inventories, and for landscape and strategic planning.

"Another aspect of the Collins Forest was the way they managed their land by a Board of Directors. When the original guy set up his will, he left just over 50% to a church, and just under 50% to his heirs. Usually, when the land passes on, the heirs feel they have to liquidate the forest to pay the inheritance tax. But the way he did it, he interjected a Board of Directors from the church that had a longer-term interest in the land.

"I started comparing that with the way the Forest Service works. The average tenure on ranger districts runs anywhere from two to seven years. That limits the extent of their professional interest and investment in that piece of ground. So how can we address investments in landscape on the order of centuries? How can we build that legacy of institutional memory?

"What we did is put together a 'Public Focus Group.' We invited people who had demonstrated over the years a personal interest in our area. We invited them to participate in setting the strategy for the district, addressing the overall landscape issues and also dealing with specific projects on the ground. We asked them to monitor our personal performance, and whether we had done what we said we would do. We tried to get a broad spectrum—industry people, small loggers, environmental groups, steelheaders, mushroom hunters, teachers. But we tried for people who would represent themselves so they don't have to go back and get an okay from their institution. We tried to avoid the hired officers who can't speak for themselves.

"When we presented this focus group idea to the Mazamas climbing club, it turned out that one of their members had been a user of the Mt. Hood National Forest for eighty years. He was ninety years old at the time. He knew stuff about the Forest that *no one* in the Forest Service knew. That's the kind of memory we need to get instilled into our management activities.

"With higher levels of the Forest Service, our program has received a curious amount of disinterest. It doesn't fit functional lines: it doesn't provide for functional career ladders, it doesn't protect functional territory and turf or functional budgets, it doesn't produce functional outputs. We have a very functionally oriented budget process. Congress appropriates us dollars for our work in very narrow areas, whether it's thinning trees, or administering timber sales, or fish projects. With our stewards, we try to comingle these funds to some extent to integrate our projects. But we can't go much further than we do without getting into trouble. We're running counter to the grain of the funding process, the promotion and benefits process, the entire structure of the organization. You may work for a line officer, but you get your promotions in your functional specialty. A wildlife biologist needs to be well thought of by other wildlife biologists at higher levels of the organization; if they want their promotion, they had better be real efficient at producing wildlife outputs with wildlife dollars. Our stewardship program doesn't feed into this organization very well, but I think it's the direction we have to take if we want to get a good look at the whole picture.

"At least there's not many people throwing rocks at us. That's largely due to my two bosses, my Forest Supervisor and Deputy. They have been extremely supportive and effective in providing a very nurturing environment for us to do this. They have made it very clear to the rest of the Forest that they want this to go ahead and give it a good chance and see what the results are.

———— 〞 ————

Many of the recent changes within the Forest Service—John Berry's stewardship program, New Perspectives, the switch to ecosystem management—occurred during a Republican administration that was not noted for its environmental concern. At least to some extent, the changes seem to go deeper than partisan politics. Still, partisan politics can play a role. During the first year of the Democratic administration in 1993, President Clinton hastened the shift toward environmental concern within the Forest Service with two dramatic moves: he appointed Jack Ward Thomas, a renowned but controversial wildlife biologist, to head the agency, and he submitted a comprehensive Forest Plan "for a sustainable economy and a sustainable environment" that he hoped would resolve the spotted owl controversy.

Option 9, as the plan is called, represents an official acceptance of "New Forestry" concepts. Staying clear of either complete preservation or full-scale commercial utilization, it establishes "late-successional reserves" in which very limited thinning or salvage logging will be permitted, provided that these activities are "neutral or beneficial" to the production or maintenance of old-growth forests. Environmentalists are disappointed that the plan does not increase the number of inviolate reserves where all logging is prohibited, but the plan does call for dramatically reduced levels of harvesting. Indeed, the official press release announcing Option 9 chastises the Forest Service severely: "Harvest levels in the President's plan take into account the fact that previous Forest Service management plans have significantly overestimated the amount of timber available for harvest every year, presenting unrealistically high harvest levels that cannot be sustained." The complaints of dissident field workers are now echoed by the President of the United States.

An intriguing component of Clinton's plan is the creation of ten "Adaptive Management Areas" in which the experimental techniques of "New Forestry" will serve to "integrate economic and ecological objectives." In these areas, commercial timber is still to be harvested, but gently: wildlife habitat, coarse woody debris, fisheries, water quality, and soil stability are to be protected, while late-successional (old-growth) forests are to be created and maintained. In another implied critique of previous practices, the plan directs the Adaptive Management Areas to move away from even-age monoculture and toward the "restoration of structural complexity and biological diversity in forests and streams that have been degraded by past management activities."

The most revolutionary aspect of the adaptive area program is its emphasis on specific locality, or the notion of "place." For years, scientists and forest practitioners have been speaking in terms of "physiographic provinces" and "watersheds" rather than political boundaries; now, this kind of thinking is reflected in institutional planning. The Adaptive Management Areas are composed of multi-owner landscapes instead of preexisting agency units. Presumably, the Forest Service will now be managing its domain in conjunction with other federal agencies, state governments, and private landowners. Decisions will therefore not be the exclusive province of Forest Service managers; local community members will help initiate and implement the projects. To move away from the regimentation and conformity required by bureaucratic structures, the plan calls for "localized, idiosyncratic" approaches that "rely on the experience and ingenuity of resource managers and communities rather than traditionally derived and tightly prescriptive approaches that are generally applied in management of forests."

The Adaptive Management Areas are to focus on processing the wood locally. By providing the maximum employment per unit of output, they will hopefully serve as prototypes for stable rural communities. Each plan is to be implemented by a local workforce, which is to receive whatever scientific and technical training it may need. Recognizing that local communities have "economies and culture long associated with utilization of forest resources," Option 9 instructs the Adaptive Management Areas to make full use of the "indigenous knowledge" of the local inhabitants: "The people have a sense of place and desire for involvement. Many of these local workers already possess the woods skills and knowledge and sense of place that make them natural participants in ecosystem-based management and monitoring."

The ideals of the Adaptive Management Areas—fostering community stability, integrating economic and ecological objectives, focusing on specific localities—are noble indeed. These high-minded phrases are not in themselves particularly new and original; the real innovation is in the realization that significant institutional changes must occur if these goals are ever to be realized. The Clinton plan states that "federal agencies are expected to use Adaptive Management Areas to explore alternative ways of doing business internally." In other words, bureaucratic procedures should be re-examined to see if they might in fact be doing more harm than good. The Forest Service policy of frequent transfers, for instance, runs counter to the notion of "place" which many people are beginning to suspect is a necessary component of good forestry. Consequently, the plan proposes to change "agency organization and personnel policies" to encourage "local longevity among the federal workforce." Similarly, the emphasis on planning often serves as an impediment to constructive action; consequently, the Adaptive Management Areas are told that "initiation of activities [should] not be delayed by requirements for comprehensive plans" other than those already required by law. The thrust of Option 9, as evidenced most pointedly in the guidelines for the Adaptive Management Areas, is clear: If forestry is to move ahead to the next step—if such ideals as "ecosystem management" and "community stability" are to be realized on the ground—business-as-usual within the federal agencies must change. The government's own impediments to good forestry must be identified and removed.

But can a presidential directive to establish ten model programs change the functioning of a well-entrenched federal bureaucracy? Political administrations last only four to eight years; the Forest Service, by contrast, has developed its own ways of doing business during almost a century of operation. The fate of the Adaptive Management Areas, as well as the impact of the rest of Clinton's plan, will be a true test of the viability—and adaptability—of public forestry.

9

PRIVATE OWNERSHIP
Freedom and Constraint

Nonindustrial private forest (NIPF) owners control 57% of the forested land in the United States, but much of this is in scattered woodlots with minimal commercial value. In terms of softwood volume, these private owners account for 30% of the national inventory.[1]

On the surface, it would seem that small private owners are likely candidates to become good stewards of the land. They can pay more attention than industrial owners to noncommercial aspects of the woods. Although the desire to make a profit might still be present, their practices do not *have* to be driven strictly by the bottom line. Precisely because they are less likely to make economically rational decisions, they are correspondingly less susceptible to the negative tug that interest-driven capitalism exerts on the forests.

NIPF owners, unlike government professionals, can practice forestry on a personal scale. Their parcels are generally small enough to permit an intimate, personal knowledge of the local idiosyncrasies of particular plots. The "stewards" on John Berry's Clackamas Ranger District practice their trade on 20,000-acre watersheds; for the Forest Service, that's considered small. NIPFs, by contrast, are often only 40 acres, and rarely over a few hundred; site-specific treatment, an important component of caring and careful forestry, is therefore more feasible. Managing Douglas-fir stands which are interplanted with green manure trees, for instance, is not very feasible when done by highly paid professionals on corporate or government land; small owners, on the other hand, can do it on their own time. They are also in a better position to perform the more labor-intensive tasks associated with hardwoods: manual release or the utilization of marginally commercial species. This is indeed fortunate, since 70% of the hardwood growing stock is found on nonindustrial private ownerships.[2]

Although the spatial scale of private owners seems appropriate to forestry, the time scale, at first glance, would appear to cause some problems. Commercial species generally take about 60 to 100 years to achieve productive maturity—roughly the life span of a human being. If a thirty-year-old person plants a tree, he/she is not very likely to be alive when it's time to harvest; if a person of any age has a chance to improve upon future generations of trees, he/she will certainly not be around to enjoy the benefits. And yet, most private owners probably feel more of a stake in future forests than distant stockholders who count their annual dividends, or even than government employees who will soon be transferred to other districts. Insofar as human beings bequeath legacies, they generally think in terms of their own families. Throughout the country, many NIPF owners have inherited land from their parents and grandparents, and they expect to leave the land to their children and grandchildren. With the continuity of the forest serving to foster the continuity of the family, these stewards have high stakes indeed in the health of the land.

Gordon Tosten

Rancher and Logger

"All my grandparents settled in the Ettersburg [California] area in the 1800s. On the Tosten side, they came out from the East in a wagon train. Ettersburg's where I was born and raised. We purchased the ranch over here [about ten miles east of Ettersburg] in 1929. Timber at that time was valueless, and a lot of the ranchers had decided that ranching was the only thing they could make money on, so they had to destroy some of the timber to make more open land. That's where the girdling came into effect. This ranch had escaped that fate. Some girdling had been done, but it proved to be too much work. Luckily for this ranch, out of the 2,900 acres, it's about half timber and half grass. There was still enough acreage in grassland to make a living on animals, without worrying about girdling trees. Nobody realized at the time that sometime in the future the timber would be worth a lot more than the grazing land.

"Actually, the timber didn't come of any value until the forties. That was when the loggers from Oregon came into this area. They had always said that the timber down here was no good. They said it was too hard, too heavy, a poorer type of Douglas-fir than they were used to cutting. But once the main cut in Oregon had been made, this timber became a little more appealing to the pocket-book. So at that time, in the middle 1940s, the timber on the ranch was sold in an entire block to a lumber company. They started their cut, but they moved out without even removing the timber they had put on the ground. They pulled out to go to a redwood area that was more profitable at that time and they never came back. After five years, my dad just called the contract null and void, and that took care of it.

"My brother and I took over from that point. We were back from the service by then and had just started our own logging operation. Luckily for us we've always had the timber on the ranch, which means we can stay right here at home and run it as a ranch and a timber operation in combination. Because without the timber operation, the ranching just won't make it. There are three part-ners—my sister, and my brother and I—and without extra capital coming in beyond what the sheep make, well, three families just couldn't live on it. Three families plus my father and mother. We have to log during the good weather, in late spring and summer, and most of the sheep work comes in fall and winter and early spring. So it gives us a year-round occupation.

"We have about 1,600 acres of timberland. It's nearly all Douglas-fir, with a small amount of redwood in one little area. We've never logged any of the redwood, except if a tree blows over we'll go in and pick it up. It's not an overripe type of timber, like some of our Douglas-fir. We've had a lot of old-growth Douglas-fir that was overripe, that needed to be harvested as rapidly as pos-sible—using common sense—to realize the best of the wood before it became more rotten. We haven't logged the young-growth Douglas-fir either, simply because it's growing every year, and that's just like interest on money in the bank. We don't harvest any of that unless it blows down or unless beetles have come in. We've had to remove some of our second growth to counteract the beetle kill. We don't spray. We have no way of fighting beetles other than removing the trees as they die. By removing them as they die, you're still making money. That's actually what most of us are trying to do anyway: make a living.

"We play the market. If this year's market says that they will take a rougher tree, a knottier tree, then that's the type we'll go after. If we have to go for the better tree because the other market is no good, then we'll go for that. You harvest what is needed.

"In years past we used to leave the severely decayed trees. We left those standing because at that time there was no market for them. We knew that someday the paper market would come to this area, and a pulp-type log would be of value. Luckily, we guessed right. Now when we take a log out that's absolutely unusable for lumber, it is usable—around $35 a thousand—for a pulp log. Every mill these days has its own chippers and its own chip bins to handle this type of log.

"Years ago a lot of loggers made the mistake of harvesting these logs, taking them to the mill, and having them culled. You'd see them piled there on every landing, beautiful logs; they looked beautiful, but they were no good simply because there was too much cull factor. They just went to waste. So one of the wise things we did was leave those trees. We could leave them because we had control of the land and didn't have to go someplace else to wait for another job. We could wait ten or fifteen years for this market, and a lot of people couldn't do that. That's where a lot of ranchers made their big mistake. They would have loggers come in and say: 'We want the entire property logged. We're in the ranching business. We're going to plant it all to grass. We want everything clearcut.'

"Looking back now, we can see that they made an awful mistake. They know they made a mistake, everybody knows they made a mistake. But it's too late. The damage has already been done. So we have a lot more erosion. They planted their grass all right, but a north hillside will not sustain grass for any length of time, if there was timber on it originally. Consequently, you come back to nothing but brush. Now, of course, a lot of people are coming back to this land, and they've got a tremendous brush problem. They're trying to compete with high tan oak and madrone brush. And this is where the spray comes in that a lot of people are against. Lucky for us, we aren't faced with that problem. We don't have to try to recover brushland to put it back in production, because we didn't create that much brush in the first place.

"A certain amount of brush is needed in order to protect little trees that are growing. So if you have a clearcut area, you allow two or three years to go by and allow the tan oak and the madrone to come, and then plant your seedlings so that they get in competition with one another. The seedling will grow faster than the tan oak and the madrone, but those trees will retain moisture in the ground

and help give some shade. You don't want any sunscald. The
seedling is growing fast, and his needles are pretty tender. He needs
some protection. So you have to have the two together, the
seedlings and the brush. Eventually the Douglas-fir will get so high
that it will overcome the brush. But a very fine line decides who's
going to outdo whom. If you don't plant seedlings quickly enough,
the brush will snuff out the Douglas-fir. But if you plant them too
quickly, you might lose them to the sun.

"And animal damage, too. Some of the animals would rather eat
a nice, juicy madrone leaf, or even oak shoots, than Douglas-fir. If
they have more of a choice, they might leave your seedlings alone.
You try to tuck your trees behind some brush where the animals
won't get to them. You plant where you think an animal can't get,
as long as you see that sunlight can get down to the little tree.
We've been planting this way, and it's paying off. We have a lot of
them now that are only three or four years old and are up to six and
seven feet tall. They were missed by the animals. The ones out in
the open that weren't missed are getting bushier.

"Of course, we won't get around to harvesting these trees for
many years. We're just now getting into our second-growth timber.
Most of the old growth has already been harvested. In the original
contract with my dad, it said that we were to remove not less than
a million and not more than two million feet a year. The problem is
that we cannot have a sustained yield with that amount of cut now
that the old growth is gone. We did that for twenty years or better
and we still have lots of timber on the ranch, but we see that we
can't sustain that kind of a cut. We're now cutting only half a million feet a year. And this year we're putting in a small sawmill, semi-
portable, to cut logs that we feel we can do better with than selling
to the mill. This will reduce our cut. If we can cut from a quarter of
a million to a half a million feet per year and saw that ourselves, I
think that we can maintain a sustained yield on 1,600 acres. [This
interview was conducted in 1980. The Tosten mill has now been
operating successfully for 14 years, cutting virtually all the lumber
harvested from the ranch.]

"We like to see trees stand just as much as our neighbors do. Our
new neighbors have told us that they like to see our trees, and they
were quite concerned when we told them we were going to cut in
certain areas. They said, 'Oh boy, here we go again. There won't be
anything left for us to look at.' Then after we cut the trees, we went
over to check about some dog problems, and they wanted to know

At home in the woods

when we were going to be through with our logging. I told them we were through. They said, 'We can't even see where you logged.' So they were pretty happy. And I'm happy, too, because I've got to live on that ranch and I like trees. My brother feels the same way and so does my dad. It's been that way all of our lives.

"I think the beauty of timber can be maintained, especially if you own your own property. Where you own the property and have control of it, you can make it as bad or as good as you want. Not everybody can be lucky enough to own land for raising both animals and timber. We just happen to be lucky. If you have a bad year in animals—coyotes or disease or severe bad weather kills off a bunch of your animals—you can say, 'Well I got logging coming this summer. I can maintain what I need.' If you own the property and if you have the right attitude about logging and ranching, you are controlling your own destiny.

"But if you go along with these preservationists who won't let you touch a thing, then you've lost control of destiny. Many laws coming in now make it more difficult and more costly for the local landowner to operate. Not that I'm saying these laws are wrong. We brought them on ourselves. I'm not sticking up for all loggers. There are a lot of bums in the past who really created a big problem. Even us. In the thirty years we've been logging, I can see a great change between the way we were doing it in the earlier times and what we're doing now. We didn't really feel it was too terrible to go through the creek to get the trees along its banks. That was an easy way to do it, and that's the way we did it. Now you find that that's not the best way. But I would hope that the laws don't get too strenuous. I think a lot of the people who are real preservationists are asking for too much.

"We're not a big corporation. On a lot of these shows, the bosses don't work. They sit in the office and they delegate their authority to another fellow who oversees somebody else. We don't do that. We run our own equipment and our own truck. We hire one faller and one fellow to help run the skidder or set chokers. I do all the cat work and my brother drives the truck. We can survive where some people in a bad year would be in real trouble.

"We have a D-6 that we used for the logging. Two years ago we bought a rubber-tired skidder, which operates much cheaper and keeps the logs clean when you bring them in to the landing. We also have a front-end loader to load the truck. And we have a Kenworth truck to take the logs to market. It's ridiculous when you think that we have somewhere around $300,000 worth of machinery and we're only taking out a half-million feet a year. The bookkeeper just pulls his hair out, because there's no way we can make it. But we make it because we have such a long, protracted contract for the future. We don't have to worry about whether we'll work next year or have to outbid somebody else. All we have to worry about is taking care of our machinery. It goes for years between overhauls. Our breakdowns are very, very few. Because we run our machinery ourselves, we know who broke it down and who to point the finger at. You see, our show is just different from the so-called gyppo or contract logger who has to put out a hundred thousand feet a day, and if he puts out ninety he's gone ten thousand feet in the hole, and that makes so many dollars in the hole. We don't have to worry about that. We take out three to seven loads a day. We take off time to manage our sheep when shearing time comes. We lay down again when we have to ship our sheep.

"We also do all our road maintenance. We have our own grader, a Caterpillar number 12 grader. We've got everything culverted and graveled. We maintain all of our roads on a year-round basis, simply because we need them all winter for checking our sheep. We run no heavy equipment on them in the wintertime. And besides the culverts, we waterbar our roads to try to stop erosion.

"Every fall we go around and open up all the culverts. Then in winter, when the first rains come, we maintain a twenty-four-hour watch, for the first few big storms, especially. That's when all those leaves wash down to the front of the culvert and plug it up. And the wind comes along, and dead twigs fall down in there. If it's a terribly hard storm at night, we go to make sure that none of the culverts is plugged up. If a culvert plugs up at eight or nine o'clock at night, the road will be gone by the next morning. So we go out at night to make absolutely sure they're open, and then we watch them all day, because we're on the ranch looking at sheep. You check them both at once. That's why you have so many erosion problems in some of these other logging areas that people love to take pictures of. It's because they weren't maintained. Nobody was out there to ditch with a shovel. After the culverts plug up, the water just goes wherever it wants.

"On some logged land, the owner lives a hundred miles away. Nobody's even there. The fellow that last logged it has nothing to do with it anymore, because he finished his job. He completed his work. He may have put a culvert in, but the owner must keep an eye on it. I don't want to say it's the hundred percent fault of anybody. Sometimes these storms will outguess you, too. You can have a crazy rain that you have little control over. While you're working on one culvert, you can have another one behind you plugging up, and before you get back to it a lot of the damage is done—especially if it's on unstable soil.

"A lot of the problems you see around here, you can't really blame on a particular individual. Just because we do it a certain way, that doesn't mean it's right for everybody. I think it's right for us. We're a set of individuals with a piece of land that's ideal. You don't even have to have a yarder on any of it, because it's not too steep. Our creeks are pretty well defined. The way it's laid out, we can log without too much damage to the creeks. And we're a family unit. Sure we have disagreement, but we can sit down and hash it over and work it out. That means a lot, too. We're real fortunate. The whole package is just right.

———— " ————

Not every logger, nor every timber owner, can be like the Tostens. Not everyone can expect to inherit a 3,000-acre ranch; and even if one did inherit a ranch, the timber might be gone and the land seriously damaged. The Tosten Ranch is the exception that proves the rule: private ownership can provide the opportunity for a personalized, caring

approach to the land; but private ownership also paves the way for rampant exploitation and the immediate liquidation of resources. The
Tostens' neighbors were unable to resist the temptation to sell off their
timber during the postwar boom. Indeed, so was Henry Tosten,
Gordon's father. If the loggers he originally sold to had not greedily
abandoned Henry's ranch for better timber elsewhere, the Tostens, like
their neighbors, would be sitting on a bunch of cutover brushland.

For those of us who have not inherited a ranch full of timber, the
opportunities to break into the business of raising trees are severely
limited. The price of country land has been greatly inflated by an
exodus from the cities. The lush forests of California and the Pacific
Northwest are now prized not only for the timber they can produce
but for the peace and tranquillity they can provide; the market price of
a plot of land is therefore determined not only by the productivity of
its soil but by its potential for housing developments and second-home
subdivisions. The Tostens probably couldn't afford to buy their own
ranch at today's market prices—unless, of course, they liquidated their
timber resources. In order to purchase land on the open market, a private buyer might well have to overcut to meet the payments—just as
MAXXAM had to do when it took over Pacific Lumber.

Theoretically, private owners have financial incentives to grow timber
as well as to cut it down, since the trees planted today will be available
for harvest in the future. According to the classical economic model,
the vested interest that private owners have in the productivity of their
land will eventually benefit society as a whole. As lumber becomes
scarce, stumpage prices will rise, and landowners will be induced to
invest in new timber production. Deforested land will be planted with
trees, while existing forests will become intensively managed. It's the
basic law of supply and demand: scarce resources will increase the
demand, and this in turn will provide an incentive for landowners to
increase the supply.

In practice, however, this classical ideal is not fully realized. The life
cycle of a commercial tree crop is several decades at least, so there's
quite a wait before the supply catches up with the demand. The self-
equilibrating mechanisms of the marketplace are geared more toward
daily fluctuations than long-term investments. Indeed, when investments require many decades to turn a profit, the law of supply and
demand tends to produce negative rather than positive results. If
stumpage prices rise today because of inadequate supply, the immediate
effect is not so much to induce investments for the distant future as to
encourage liquidation of present inventories. When the price of timber

doubles within a year (as it has been known to do), the first impulse of the small landowner is not always to plant another forty acres of seedlings; instead, it is to harvest the timber he already has. Thus, his timber is added to the presently available supply but is effectively removed from tomorrow's market. The timber shortage is alleviated in the short run but aggravated in the long run.

Even as stumpage prices reach astronomical levels, the small private landowner shows little interest in tying up capital in reforestation and timber stand improvement. When investments are made, they are often subsidized by the government. Since 1974, the Forestry Incentives Program has offered to pay a considerable proportion of the costs— sometimes as high as 75%—for reforestation and timber stand improvement. (In California, a state program was also available during the 1980s, but the funding has since disappeared.) The public sector has helped to underwrite investments in tomorrow's timber only because individual landowners working within a free-enterprise economy have been unwilling to make the investments on their own. A survey conducted by the Forest Service revealed that 80% of the private landowners intended to harvest their timber without making invest- ments in interim management, while only 5% showed any interest in improving their stands prior to economic maturity. (The remaining 15% were holding their land for nontimber purposes, such as recre- ation, speculation, or development.) The study concluded: "Most forest owners do not consider timber-growing investments to be suffi- ciently profitable to take priority over other investment or consump- tion opportunities."[3]

The relationship between the small, private landowner and the gov- ernment has become increasingly complex in recent years. On the one hand, the government offers the carrot: shared costs for investments in tomorrow's timber. But the government also wields the stick: restric- tions of a person's freedom to manage his own land. Private owners such as Gordon Tosten believe in "controlling your own destiny." This sounds enviable, but since the possibilities for exploitation are always present, the owner's control is circumscribed by a myriad of regula- tions intended to protect the land from abuse by its master. This con- flict—owner control versus environmental restrictions—is a strong undercurrent in timber politics. In the context of the damage that can be (and often is) inflicted upon the earth in order to realize a quick profit, the restrictions seem quite necessary; but in the context of American individualism, the restrictions seem like an extreme invasion of privacy—and they burden the owner with a considerable expense.

Wayne Miller

Nonindustrial Timberland Owner

"I used to be a photographer, and in 1952 I did a small story for *Life* magazine up in Washington on Crown Zellerbach land. I was quite taken by their young, healthy, beautiful trees. Joan got excited about it too, so we spent the next six years learning what we could about forestry. We learned how to read aerial photographs, we learned that we needed good soils, and that we had to protect ourselves from fire, insects, and people pressures. We also learned that when you buy your land, make sure it's surrounded by a large timber company, because they'll fight your fires for you.

"Eventually, we located our property north of Fort Bragg. It wasn't considered of much value at the time because it was brush land, it had been heavily cut over. We bought it with the idea in mind of enjoying the land, raising trees. It was 1200 and some acres at the time. Since then, we've added to it. Now we have about 1850 acres.

"With a Registered Professional Forester, we developed a management plan, breaking the land up into eight or ten separate units. We figured we could go back into each of these units every ten or fifteen years and selectively harvest about thirty percent of the volume. We're still following that original program. We cut about thirty percent if the market's good; if the market isn't good, we don't cut. We gave it some time, from '58 to '72, before we started harvesting. Even since '72, we've missed some years. Now, it couldn't be coming along better.

"Those of us who are involved in nonindustrial forest land are individuals, not corporations. The average owner is middle-aged or older. We have a limited life span. If we want to harvest the trees we plant, we have to wait fifty, sixty, seventy years. So we have to think about our families, how to transfer our excitement about the property and its ownership to our children. When we die, we want them to hold on to our dreams, our goals for developing a beautiful and productive forest for generations to come.

"But it's becoming harder and harder to maintain our enthusiasm, because the increasing forest practice regulations have created excessively high timber harvest plan costs. You may assume that you can harvest your trees, but you may not be able to afford it. Costs of getting a timber harvest plan approved, under the new rules being proposed by the State Board of Forestry, will be ten times greater than in 1989. In the coast district, if you owned less than 20 acres in 1989, it cost you $450 for a timber harvest plan. Under the new rules, it could cost $4,900 dollars. If you owned less than 640 acres, it was $1,500 in 1989; with the new rules it could cost $17,720. For over 640 acres, it was $2,269; under the new rules, it could be $22,500. And even these could be higher if you are required to provide additional information such as watershed and wildlife studies.

"Three professional foresters supplied these figures, which they would give as bids to prepare the plans for their clients under these proposed new rules. And the rules are as complicated as they are expensive. The estimated costs would include describing the silvicultural practices, creation of an inventory and growth data base, doing a study to prepare a sustained-yield analysis, developing an estimated harvest scheduling for a hundred year horizon. For the maintenance of the older tree requirement or late seral stage, you might need to hire a wildlife biologist. Domestic water sources, one thousand feet downstream from the site, can cause real people problems and expensive studies. If you're in a sensitive watershed, mitigation measures may be defined by political, not biological, criteria.

"Timber harvest plans require consideration of not only the scientific aspects but the social aspects. This is quite proper, but at times it can become confusing and frustrating. We are required to protect the forest's 'aesthetic values,' even though the phrase 'aesthetic values' is nowhere defined in the law. If we just had to deal with science, it would be comparatively simple, but subjective regulations are hard to satisfy and political pressures usually prevail.

"Strangely enough, our forest regulations can result in poor forestry. With timber harvest plans so costly, you can't afford to take out only a few trees or to do the needed thinning of that little patch over there. You just can't get a timber harvest plan every time you want to do a little work on your land. And when you do get a plan, economics may force you to cut as much as you can. This doesn't promote good stewardship.

"For some forest owners, there is a way to avoid these continuing costs. If you own less than 2500 acres of timberland and you plan on harvesting parts of the forest every ten or fifteen years, then you can get some relief from these expensive regulations. I encourage small landowners who want to make repeated entries to make use of the state-created Nonindustrial Timber Management Plan (NTMP). Once your NTMP plan is approved, you won't have to apply for a timber harvest plan every time you wish to harvest. You just notify the California Department of Forestry, telling them that you plan to harvest as stated in your NTMP; then you can start work immediately. This means that you can make those frequent re-entries and can take advantage of the upticks in the timber market. It means that you can make some long-term management and investment decisions. It provides the only security available in today's regulatory climate. Under these proposed new rules, the costs of an NTMP on smaller ownerships may not be much more that the cost of a single timber harvest plan. Best of all, it goes with the land and remains in force as long as you follow the original plan. Even though all laws are vulnerable to public pressures, presently it's our best answer.

"Still, I remain concerned with ever increasing and changing forest regulations. They've taken a lot of the fun out of forestry for me. I have more questions than I have answers. To what degree will I have control over the future management of my land? Should I plant a tree today if changing regulations may not allow me to cut it in the future? Will I be able to afford increasing regulatory costs? What happens when I die? Will the State allow my family to cut enough trees to pay the estate taxes, or will my family have to sell our land and trees to industry or to subdividers? These are real questions for which there are no real answers.

"The pressure of these ever changing regulations, and the constant threat of having more regulations, results in bad forestry. It causes us *not* to make long-term forest plans and investments. These rules designed to protect the forest environment lead to quite the opposite results. We're destroying the forest base which, as individuals, we want to embrace and protect.

"New laws and regulations are not the answer. To assume that bad forestry can be regulated out of existence—just by passing new laws—is naive. We should step up the enforcement of the existing laws and regulations rather than look for new ones.

"We must find ways to encourage good forestry, not discourage it. I've been at social gatherings where I'm asked, 'Well, what is your interest? What do you do?' Some people say a doctor or accountant, and I say, 'Well, I'm interested in forestry.' Invariably they say, 'You don't mean you *cut trees!*' That's not very encouraging; that's a disincentive. As individuals, we like to think we're

Douglas-fir: a wise investment?

caretakers of the Garden of Eden and that we're doing the best job possible. We should be encouraged to take care of forest land for the public good. Not many people realize that over 50% of the private timber and timberland in California is owned not by big companies but by the concerned small forest owner.

"I've been racking my brains for several years to figure out how to make the growing of trees seem admirable, a desirable way of life. People should be saying, 'Gee, I wish I could be doing that too, taking care of the land, the trees, the wildlife, and, at the same time, producing forest products.'

"Forestry is an act of love, an act of caring. But with the excessive regulations and disincentives, love can disappear, leaving behind a forest that is a stranger. I'm fighting to protect my love affair because I believe this is the best way to create a healthy, productive forest.

——————— ” ———————

The Politics of Regulation

There is an implicit assumption in the body politic that all problems can be solved if only the proper laws are passed. In these days of heightened environmental awareness, this assumption has led to the imposition of increasingly strict regulations to prevent abuses by timber owners and logging operators in both the public and private sectors. But do the laws work? Can regulations and restrictions guarantee good forestry?

On a strictly personal level, the average worker in the woods feels little sympathy for the regulations that are supposed to govern his behavior. Indeed, most loggers are intensely alienated by external restrictions. There is a mystique of freedom about the woods, a mystique shared by the logger, the mountain climber, the fisherman, and all other men and women of the outdoors. To be working in the woods, to *know* the woods, and yet to have one's actions dictated by office-bound bureaucrats and politicians hundreds of miles away seems the height of folly. The regulations are commonly experienced as personal insults, threats to the personal integrity of the people who work on the ground.

When a fundamental antagonism exists between the regulators and the regulated, the mechanics of enforcement become rather tricky. Strict and precise standards of behavior must be explicitly stated. Money must be allocated for enforcement officers to conduct on-site inspections. And there must be severe sanctions to effectively prohibit violations of the law. For the regulations to work, all three of these conditions must be met: precise standards, frequent inspections, and prohibitive sanctions. Do the current forest practice regulations effectively embody these three elements?

California has one of the strictest sets of forest practice laws in the nation. If the laws are to work anywhere, they should work here. In the early 1970s, the California Supreme Court ruled that the existing forest practice laws had to be overhauled because they were drawn up and enforced by a Board of Forestry that was dominated by the timber industry. The new Board of Forestry had a more balanced composition and proceeded to draft standards that, on the surface, seemed precise and explicit.

The new board defined the different silvicultural systems quantitatively, requiring that a certain number of trees of a given size be left standing for a selective cut, another number for a shelterwood cut, and so on. It required that a specific percentage of the forest canopy be retained within a set distance from streambanks. Utilizing maps of the entire forested region within the state, it labeled each piece of land with a specific slope and soil type and computed a numerical "erosion hazard rating," which could supposedly predict the extent of erosion to be expected in the wake of logging operations. Armed with all this data, the Department of Forestry could presumably require preventative measures to alleviate adverse effects on the environment.

In practice, however, the use of explicit numbers on a piece of paper did not result in the application of precise standards for real-life situations. Despite the quantitative definitions of selective cutting, shelterwood cutting, and so on, many logging operations were approved by state officials under the euphemistic name of "overstory removal," a term that was not written into the law and had no legal definition. Despite the requirement that 50% of the streambank vegetation be saved, inspectors and loggers alike could only guess at the amount of shade preserved, and the requirement only pertained to those streams that had found their way onto official Geodetic Survey maps.

Indeed, the reliance on mapping proved to be a weakness of the regulatory system. Streams were not defined by their actual aquatic properties, but by whether surveyors in years past had deemed them significant enough to be officially registered on the maps. Specific slopes and soil types were numerated to describe every square inch of forested land, but in fact the maps only represented approximate averages: the mapping units were composed of ten- to forty-acre minimums, and within each unit the slope and soil types often varied considerably. A forty-acre unit that was registered in the 31 to 50% slope class might have contained ten acres with a 70% slope and another ten acres with only a 20% slope. A fifteen-acre unit mapped as having a stable type of soil might have contained several acres of an unstable soil type that never appeared on the map.

To some extent, these imperfections in the law can be remedied by breaking things down into smaller and smaller units. Imprecision, however, is inherent in the regulatory process. The regulatory bodies cannot be expected to know each and every locality intimately and specifically. They are forced by the very scope of their operations to rely on imprecise and incomplete information. This causes the science of logging regulation to be in fact a pseudoscience.

The Department of Forestry, for instance, may grant permission to log a given area with tractors because it has labeled the area with a low erosion hazard rating. The rating is an actual number, lending a false feeling of objectivity and precision to the entire process. How effective are these numbers in predicting the actual erosion caused by logging? A study of 27 logging operations in California compared predictions from the erosion hazard ratings with the amount of erosion that actually occurred. In only 7 cases did the predictions even approximate the actual extent of erosion. In 4 cases the predictions grossly overestimated the erosion, while in 16 cases the rating system grossly underestimated the total extent of erosion.[4] A rating system as ineffectual as this is hardly a scientific basis for precise and effective regulation.

Recognizing the inherent difficulty in stating standards applicable to each and every circumstance, the writers of the California forest practice rules qualified many of their regulations with "weasel words" (as they are known in the trade): "reasonable," "unreasonable," "minimal damage," "in the best judgment of," and so on. For many of the most important regulations, authority was thereby shifted from the rules themselves to the discretionary powers of the inspectors. As any good bureaucrat knows, consistency cannot be purchased at the expense of reasonable flexibility. Since forestry is as much an art as a science, a good case can be made for the need to bend the laws to reflect the natural variations in real-life experiences; but the substitution of discretionary power for written authority tends to take the teeth out of the laws themselves, for there are always plausible reasons for making exceptions.

A classic example of balancing strict-sounding verbiage with weasel words occurred when the Board of Forestry responded to the court-ordered mandate to consider the "cumulative impacts" of past and present logging operations. Timber harvest plans, said the Board, were to watch for several possible cumulative impacts: stream sedimentation, high water temperature and dissolved oxygen, loss of topsoil; they were also supposed to pay some attention to the preservation of snags, woody debris, and a multi-story canopy for wildlife. But what exactly were they to *do* about all these factors? The wording in the rules is interesting: the "guidelines *may* be used" when evaluating cumulative impact. "No actual measurements are intended."[5]

Even if the government could successfully codify and quantify the regulations, it would still face the problem of on-site enforcement. The difficulties of enforcement in any industry are proportional to the degree of decentralization, and logging is perhaps the most decentralized of all industries. Inspectors cannot satisfy their obligation by visiting a centralized facility; instead, they must sally forth into the woods to check up on logging shows that are incessantly on the move. The inspectors are supposed to examine each timber harvest plan before it is approved by the state, and they are expected to visit the site again after the operations have been concluded. But to enforce the rules rigorously, they would have to remain on the job when the actual work is being performed, and this would require a veritable army of inspectors. Since the inspectors are government workers, the public is presented with a difficult decision: do we sustain a tax-supported superstructure of nonproductive employees (inspectors), or do we settle for occasional spot checks and incomplete enforcement of the laws?

The final requirement for workable regulations is a set of prohibitive sanctions. Penalties must be established that make it either costly or painful to disobey the law. A gentle slap on the wrist, a mere embarrassment, cannot in itself produce upright citizens; culprits must be made to feel that crime doesn't pay. But is disobedience of a forestry regulation really a "crime"? Very rarely do loggers or landowners wind up in some overcrowded prison for cutting timber too close to a stream. Most state laws provide for up to a year in prison or a $1,000 fine for an infraction of the rules. But the jail terms are little more than idle threats, while $1,000 fines do not function as meaningful deterrents to people who can make ten times that much by cutting a single old-growth tree. In California, an errant timber operator can also have his license revoked, but this too is rarely invoked; in the first ten years following the new forest practice regulations, only two licenses were taken away—and one of these was returned upon appeal.[6]

The only club with any clout is the provision that offenders will be held responsible for the reparation of all damages caused by a violation of the rules. This could run into money, but it is virtually impossible to implement. How does a logger replace the earth that has been lost to erosion? How can he replace an archeological site that he has inadvertently destroyed? In practice, this provision is most commonly enforced merely by requiring the offenders to take more precautionary measures in their future operations. A little good will might be lost by an infraction of the rules, but the show must go on, and business will continue as usual.

With full-scale enforcement not feasible and effective sanctions untenable, more and more emphasis is placed on the first step of the regulatory process: devising the rules. Rules can be made inside conference rooms and office buildings, far away from the woods. They are circumscribed and finite, far easier to control and manipulate than the forest itself. They have a seductive appeal: once a rule is passed, we have an illusion that the problem has been solved. Unfortunately, real-life doesn't always work that way. Debates over the rules might grab our attention, but new rules, in and of themselves, don't necessarily lead to sweeping changes in forest practices.

Although rule making should not be confused with the actual practice of forestry, regulatory politics have generated some beneficial results. Forest practice restrictions have alleviated many of the extreme mal-practices that were common in the early days of cat/truck logging: streams are no longer made into skid roads, tractors no longer operate on some of the most unstable slopes, and the forest is no longer lost and neglected after the first harvest. Although enforcement is not universal, the regulations still serve as educational tools. Today, most loggers know they should regenerate the areas they cut, even if the inspectors are unable to count each and every tree. And the landowners are told that regeneration is a mandated rather than discretionary expense; if left to the whim of individuals and the thrust of interest-driven economics, replanting would not be as widespread as it is today. Many professional foresters are explicitly thankful for the regulations; now, instead of trying to talk their clients into practicing good forestry, all they have to do is point to the rules.

Some regulation is clearly necessary to curb abuse, but are there other forms which the regulations might take? Does the regulatory system always have to lead to alienation among the people of the woods? Could we possibly devise some structure whereby the *making* of the rules would not be so divorced from their *enforcement*?

Henry Vaux

Former Chairman, California Board of Forestry

"I graduated from college in the midst of the Great Depression in 1933, and that wasn't a very good time to get a job anywhere. The New Deal was just in formation, and that made forestry somewhat attractive. I came out here to Berkeley to go to forestry school, and outside of World War II and a few teaching stints at Oregon State, I've been here ever since.

"I was appointed to the California Board of Forestry in 1976 and served a little less than seven years. I was one of several public mem-bers, which meant without any pecuniary interest, and I was also the Chairman. One of the remarkable things about the Board of Forestry is that it's an unpaid position. You get your expenses paid, something like $50 for every day of meetings you attend. But nobody ever made a lot of money serving on the board, or even a living wage. That has to be remembered when people criticize the board members; these are people who are donating a lot of time and energy to public service.

"The first regulatory act in California was passed about 1945 or '46. The original Forest Practices Act, the only people really interested in it were the industry people. They didn't actually write it, but they were very heavily involved in helping to write it, and it had many of their ideas. That act, as you recall, was declared unconstitutional in 1972 because the landowners were in control of making the rules through district forest practices committees. On the Board of Forestry there was one livestock representative and a water representative, but nobody who would now be considered as representing the public interest, with the possible exception of the chairman of the board, who was appointed by the governor. There were several representatives of timber interests, so it was essentially perceived as industry self-regulation.

"When the act was thrown out, the state was left without any regulation at all. The big question was: what would come next? There were two rival bills. My first association with it came when Assemblyman Z'Berg contracted with the Institute of Ecology at Davis to propose a new law which he could introduce. I served on a committee appointed by the Institute to draft a piece of legislation which became the original Z'Berg bill. It had no provision in it for rules of forest practices, and the enforcement mechanism was quite different from what was eventually adopted. It provided that there be a special examination of eligibility for foresters who wished to become timber harvest planners. These people would give specifications of what the logger ought to do on the ground to achieve the objectives of the act. The review would come only on the completion of the harvest, to see whether the objectives of the act had been achieved. If the plan wasn't followed, action would be taken against the logger, or if the plan itself wasn't effective, action would be taken against the license of the certified forest planner. So it was a different approach to the enforcement problem, and one that didn't rely on a large book of specific rules that you had to comply with. The foresters only had to achieve the objectives of the act:

continuous forest production, adequate restocking, prevention of erosion, maintain water quality, reasonable attention to scenic values (although that became a very hard thing to set standards for).

"When the bill was first introduced, everybody opposed it except the Sierra Club. The opposition came from both the landowners and the professional forester's groups, who objected to placing this responsibility on the forester. The foresters saw it as a double jeopardy, as having two masters: both the public interest and the owner. I always thought that was a red herring, since many people have dual responsibilities; the most obvious example is a certified public accountant, who has an obligation both to his employer and to the public interest.

"At the same time in the Senate, there had been a bill introduced by Nejedly which followed the format of the original Forest Practice Act, with rules and regulations defining everything. The difference was that the board was responsible for adopting those rules rather than the district committees of landowners, and the board was reorganized to have more public interest members. In the political resolution between the two different bills, about the only thing left from the old Z'Berg bill was the timber harvest plan, but that was changed to fulfill a completely different function.

"The major defects of Nejedly–Z'Berg have become obvious over time. Once you set up a rule-making machinery, then people seem to run wild in trying to regulate everything, whereas in the nature of actual forest practices, you simply can't make enough rules to cover everything that's going to happen, to provide for all the variability. There are so many sources of variability out there— the soil, the season, the vegetation, the landscape—that it's hard for me to visualize how you can ultimately get away from using the brains of the individual on the ground who knows the land, knows the processes, and has a good commonsense understanding of what the objectives are. I don't think there's a substitute for the human brain in a situation as complicated as the application of forestry.

"One of the big problems with the rule-making approach is that once established, it becomes an attractive device for carrying out other agendas besides the ones which the rules were originally designed to embrace. We didn't realize it at the time, but forestry, because of it's multiple-use aspects, is a natural for laying layer upon layer upon layer. Everybody is trying to make use of the same channels.

"I think for awhile the new act worked reasonably well in bringing about some improvement, but when the intensity of the political pressures increased greatly, the structure gradually became very cumbersome. The board's agenda began to be controlled by external events. The first thing that came along was the Redwood Park expansion. The proponents of park expansion were trying to use the board's process as a way of preempting logging in areas that were being considered for addition to the park. Every timber harvest plan that came to the board was challenged, and most of them eventually made their way into court. It was part of a political agenda, a delaying mechanism to stop the logging until the land could be acquired for the park. The law was not originally intended to be used that way.

"The board was constantly at loggerheads, divided into two camps—industry versus the environmentalists. For about two or three years, people on both sides seemed to make a distinct effort to be cooperative. Then after about three years, there came to be change in industry perceptions. Instead of having an 8 to 1 or 7 to 2 vote, it was 6 to 3 or 5 to 4 on almost all actions, according to straight party lines. The industry's party line, in particular, was becoming more doctrinaire. This was disappointing. I guess if I was the kind of person who got easily frustrated, I might have had sense enough not to take the job to begin with. I sometimes think I should have been more irritated. But I did feel relieved to get off; seven years was enough to put up with that kind of monkey business.

"In saying what I have about the regulatory system, I don't mean to imply that I'm against it. If you jacked up the existing forest practice act and drove a new system under it with the same policy objectives, that would be ideal—an act to accomplish the same ends, but by modified means. When I criticize the regulations, it's because, as anybody can see, they have weaknesses. The question is one of judgment: when can you come up with something better? I think a lot more thought ought to be given to a comprehensive review of the *structure* of the system, as opposed to changing just this rule or that rule. In terms of systems theory, somebody ought to examine: is this really a 'system,' or is it a hodgepodge which, because of historical pressures over time, has been mangled to the point that it no longer has adequate properties as a system? There ought to be a certain internal logic to it; there shouldn't be too many unidentified conflicts within it. A system is a logical construct with internal feedback systems that bring attention to problems and start corrective actions, and that isn't the way the forest practice rules were built over the years. They were developed more ad hoc, a series of separate responses to external pressures.

"I think we've come to the point now that most people are ready to look for alternatives. I have always had in mind that the way to regulate something like this is to have a professional do it on the ground, where he can weigh and evaluate a whole lot of things in terms of his judgment. Then you can approach enforcement in terms of sanctions against those individuals, like malpractice against doctors. The problem is, you can't do that unless you have the cooperation and the concurrence of the profession involved. The profession of forestry had a wonderful chance to embrace some real responsibility, but the profession as a whole was reticent to embrace it. It may be too late to embrace it now, in this age of sweeping litigation.

"I also think the use of incentives of one kind or another, through taxes or subsidies, ought to be studied much more carefully. In principle, I think incentives are fine, but there's always a big jump between principle and practice. People have always been shy of the subsidy thing, because there's a general feeling that subsidies are bad. But I think they ought to be much more widely considered, and probably used, in many environmental matters, because the trouble with environmental values is that they often don't show up in the economic system at all, and so one way of giving them value in an economic sense—'the pricing of extra-market benefits,' as I used to call it—is to use subsidies.

"I'm intrigued by the fact that they are now trying to lessen the burden on the small forest owner. When the Z'Berg act was originally passed, my observation at the time was that even though this is eventually going to put the small owner out of business, something like this has to be done in order to protect the value of the forests. So I supported the act, but with reservations. The only thing that's kept the forest owners in business has been the simultaneous meteoric rise in stumpage values. The job got tougher but the rewards got greater with the passage of time. But that will inevitably reach a limit, so I think the board is on the right track in trying to find different ways of treating the small forest owner.

"I don't see why they couldn't try, with forest owners of 100 acres or less, this business of having a forester who would be willing to assume this responsibility. Make it voluntary between the owner and the forester. Get some experience with people who are willing to take some responsibility; give them an opportunity to do it that way. That way you limit your risk. I think this approach would work well with other ideas as well: try it out on a small scale and see how it works. One of the problems with the rule-making system is that you have to jump off a sixty-foot diving board right off the bat instead of sitting with your toe in the water and finding out how it feels.

"With the rule-making system, we try to make a uniform prescription for the whole state of California. Talk about diversity, California's got it—not just in biological diversity, but it's got a great diversity of people. That's why I think we have to approach it not by trying to design a system that's good everywhere, but design a system that will work but that has lots of room for change as you go across this diversity.

———— ,, ————

The regulatory procedure, for better or worse, is an inherent feature of private ownership. There is a basic paradox in the ownership of forest resources: if you own the forest, you possess the means to exploit it, yet if you don't own the forest, why should you care enough about its future to do it justice? Forest practice regulations are intended to restrict the possibilities of exploitation—but do they simultaneously increase the degree of alienation to the point where the owners no longer really care about their land?

Some flexibility must be built into the regulatory system to keep it from enslaving us all. Infringements upon personal freedom must be weighed against environmental considerations. In California, the law takes mercy on the truly small landowners who wish to harvest trees on less than three acres: instead of filing an elaborate and costly timber harvest plan, all they need is a one-page exemption form which is not subject to review. It sounds benign, but this exception has provided a convenient loophole for bypassing stringent regulations. In Humboldt County alone, THP exemptions increased from 125 in 1987 to 397 in 1992—and many of these were actually used for commercial operations.[7] In one case, two partners bought a six-acre parcel, divided it down the middle, filed their exemptions, and hired a logger to cut both halves simultaneously. In this manner, a six-acre clearcut was achieved without any serious regulatory constraints.

Should we tighten up the rules or not? Should the owner of a one-half acre parcel have to file a timber harvest plan if he or she wants to cut down a single tree to let sun into the yard? It's hard to say. Unfortunately, rules do not always provide the answers, nor do they always produce the intended results. When Jan Iris was helping the forests by thinning brush and reducing the fuel load, he was granted THP exemptions because he did not remove wood that was deemed of commercial value. Through his pioneering efforts, however, he managed to make enough money by milling tan oak to pay for his work in the woods. But as soon as he had shown that tan oak could be of commercial value, he was no longer eligible for exemptions—and the costs of filing THPs in order to heal the forest were prohibitive. The rules we make, for all our best intentions, can impede as well as promote good forestry.

The problems of regulating private land stem in part from the perplexing relationship between the individual and the state. The private owner who possesses the means to damage the environment must certainly be controlled, but this control by the state now seems out of control. The only way to resolve this paradox is to redefine the concept of "ownership," moving it more toward an idea of "stewardship." A steward is a caretaker of the land, a person whose personal or professional satisfaction comes from keeping the forest healthy and productive. Some private owners function as stewards, while others do not. The purpose of regulations is to turn *all* owners into stewards, at least in a functional sense. Unfortunately, real stewardship entails a relationship with the land that is not easily instilled by force.

PART IV

PEOPLE IN THE FOREST

Are We Sustainable?

10

COMMUNITY INSTABILITY
Jobs, Owls, and Machines

Economic and political forces clearly have significant impacts upon the forests. But as we change the forests, they in turn create repercussions within our human communities. In the Pacific Northwest, the harvesting and processing of forest products have historically accounted for approximately one-fourth of the total manufacturing employment.[1] In 17 of Oregon's 36 counties, forest products have accounted for over half the manufacturing employment.[2] What happens to all these jobs when the available resource base begins to disappear? And what happens to the forest-dependent communities?

Mill closures in recent years have devastated the social landscape in the Pacific Northwest. In Oregon, over half the closures have occurred in communities with less than 5,000 people.[3] With the loss of their major industry, these small towns are literally falling apart. High school graduates, unable to find jobs, flock to the military; older workers, also unable to find alternative employment, drop out of the work force altogether. With depressed local economies, the decline in tax revenues cripples governmental services right at the point when they are most needed. In Oakridge, Oregon, property tax revenues declined by 42% in just four years. To make up the difference, local leaders placed seven different tax levies on the ballot—but all of them failed. Those who could moved away; those too old stayed behind, resigned to finishing out their lives midst a declining social and economic environment.[4]

Why? Who is to blame? What economic or political forces have wreaked such havoc?

The trouble seems to have started back in 1979, when the Federal Reserve tightened the money supply in order to curb double-digit inflation. As interest rates soared and home construction came almost to a standstill, lumber and plywood prices tumbled. By 1982, 118 mills had closed down in Oregon and Washington,[5] causing severe

Ghost town

unemployment in timber-dependent communities. During 1981–82, Coos, Douglas, and Lane Counties experienced a population loss of over 7,000 as people left their homes in search of work elsewhere.[6] The impact of financial policies made in the nation's capital was felt in small logging towns three thousand miles away.

By 1986 lumber prices had stabilized, but employment continued well below the level of the 1970s. In fact, during the late '80s and early '90s, prices soared to historic highs—yet people were still out of work. Clearly, there was more amiss than fluctuations in the business cycle.

Today, any mention of unemployment in the timber industry is immediately associated in the public mind with owls. The northern spotted owl, according to wildlife biologists, is an "indicator species" for the health of old-growth ecosystems. These owls like to nest in the tops of snags, and they like to eat such critters as the red-backed vole, which itself is dependent on fungi found only in old-growth forests. If the old-growth forests disappear, scientists worry that the owl and many other wildlife species will also disappear. Rather than inventory every threatened red-backed vole and western big-eared bat, biologists reason that the survival of the owl would constitute sufficient evidence that the old-growth ecosystem is alive and well. Based on the legally mandated need to protect wildlife diversity, Federal Judge William Dwyer issued a moratorium on old-growth timber harvesting in several National Forests until the Forest Service could come up with a plan to protect the habitat of the northern spotted owl. (For a brief background of the court decisions, see Chapter 8.)

Protection of the owl, however desirable from an environmental point of view, has had sweeping social. and economic implications. Because the vast majority of large timber on private lands has already been cut, most of the remaining old-growth stands are found in the National Forests that are affected by Judge Dwyer's rulings. Almost half of these ancient forests—about 2.3 million acres in Washington, Oregon, and Northern California—had been slated for timber production until the courts stepped in.[7] With these lands removed from production, a significant portion of the resource base was lost to the timber industry. And when old-growth harvesting from the National Forests came to a virtual standstill, thousands of timber fallers, truck drivers, and mill workers suddenly found themselves out of work—soon to be followed by car salesmen and grocery clerks and schoolteachers in timber-dependent communities. Estimates of the total employment decline in the Northwest caused by the protection of spotted owl habitat range from 15,000 to 34,000.[8]

Randy Spanfellner

Former Sawyer

"I was mainly in the sawmill end of it. I started there when I was in high school back in 1973. I worked full time when I was still in school, pulling green chain on swing shift. My dad was a carpenter, and he fell off a roof and broke his back—so I had to go to work to help support the family. I started out making $4.20 an hour, and I thought, 'Man, this is great.'

"I worked there for twenty years. I had probably every job in the sawmill. I ended up as a head sawyer. I sawed all the grade out of the logs. You need a lot of knowledge and coordination to run a headrig—you have to run a log back and forth and turn it at the same time. It takes years to get proficient at it.

"We cut old-growth logs, no second-growth at all. We cut twelve-inch clears that we would export to the Japanese. When you get big old-growth logs, you can go in twelve inches deep on all four sides and it will be solid clear. Those are nonexistent now. They won't let us cut anymore. That's why we lost our jobs.

"We were probably the number one mill in the world as far as cutting clears. Then the environmentalists came into the picture about four years ago. They didn't want us to cut any more logs, and what they used was the spotted owl as their excuse, that the spotted owl was going to be extinct and they only lived in old-growth forests. Well, that's a matter of opinion, but I guess they've got enough strength that they can just put all these stops on logging.

"We had no sales because they were all tied up in court. These environmentalists would go out with lock and key and chain themselves to trees and drive spikes into the trees. Our loggers would go in with their trucks to bring them out and they couldn't even get through. Well, the federal government seen all this happening and they would lock these sales up and they wouldn't let us take the logs out.

"I hit spikes that were driven in trees. We were using big band saws. I'd sit in a little cab, and I had a big carriage and I'd run these logs back and forth in front of me sawing them up. Boy, you hit a spike and it tears all the teeth out of the saw. It could even tear the saw off the wheel, and it could've killed somebody. I had an off-bearer that stands at the head rig and directs the lumber off the saw—that guy's got his life in his hands with people spiking these logs. He's probably about four or five feet from the saw. We've had saws come off and wrap around off-bearers. Luckily none of them got hurt too bad.

"When the logs started running out, they put us on four days a week. Then after awhile they laid off the night shift. Then it went to where they were laying the rest of us off for a week, two weeks at a time. They'd shut the whole plant down for awhile, then start back up. Last January we got a letter one day saying March 9th was our last day. They gave us a sixty-day mandatory notice: they were going to shut her down. And that's what they did. We sawed up the last logs and they shut the doors. Come March 9th, we had a big meeting out in the parking lot and he gave us all a 'Dear John' letter and a hundred dollar bill. That was it. Twenty years there, I got a hundred dollar bill.

"Everybody was just wondering what they're going to do. Out here it's mainly all logging and sawmills. A lot of the other mills went down before us, so any of the jobs to be had were already taken. We seemed to hang in there a little longer than some of them. My brother, he drives a log truck, and he's still working. He works on small second growth; he takes 'em off farmers' patches and private property owners. There's still a few mills cutting studs.

"I've always bought and sold a little real estate, so I went ahead and got my real estate license when the mill shut down. A lot of the other guys, they're in deep trouble. They've got this program where they'll retrain you while you draw unemployment, but most of the guys cannot afford to take advantage of the training because they cannot live on unemployment. That forces them, rather than get retrained, to go out and take any job they can get—as long as it makes more than unemployment. So a lot of them just took manual jobs, whatever they could find. I know lots of guys that've lost their cars. Had to give them back.

"This whole experience was pretty devastating for my family. Every two weeks I had a check for twenty years, and all of a sudden I didn't. Just knowing that it was coming was real stressful. My wife used to work part-time, and she had to go full time. She's an office manager at a dental office. She didn't want to have to work full time. (I hear about it almost every day.) We've got a little boy at home who's four, and another son who's twelve. She doesn't like to have to go off every day and leave them all day long.

"My four-year-old is too young to notice, but the twelve-year-old knows we're broke all the time. He understands, he's doing pretty good. He makes his own money. Today he's selling hot dogs at a rummage sale; he feeds the neighbor's dogs; he mows grass, cuts firewood.

"Starting in September, I'm going to go to school at this retraining program. I'll get a degree as a manufacturing engineer, a two-year associate degree. I wanted to have a trade where I can get a job. So many of these guys go to school, and then they can't find a job. I don't really care to be a manufacturer; it's not really what I wanted to do, but I want to be employable. I'll do the best I can. Maybe I can be a computer robotics engineer.

"Personally, I don't really blame the environmentalists for what happened—except for the tree spikers. What we needed was a happy medium. If the sawmill owners kept going the way they were, running two and three shifts a day, they would have cut every single log in Oregon and reaped all the profits. It's not like they passed it on to the workers for any type of profit sharing or retirement. When the logs were gone, the owners would just have walked away. I feel these people just stopped it while we still had some trees. I'm an outdoorsman, I love to hunt and fish—so I'm glad to have some trees. It just happened to be my job.

——————— *,,* ———————

Mechanization: John Henry Can't Compete

The protection of spotted owl habitat is not the only cause of unemployment—nor even the largest one. For the past thirty years, numerous studies by a host of separate agencies have been predicting a serious loss of jobs—ranging from 13.4 to 38.5%—in the Northwest timber industry before the year 2000.[9] Indeed, the decline in employment was well under way long before the spotted owl controversy tied up over two million acres of federal land. From the mid-sixties to the mid-eighties, employment in Oregon sawmills declined by 30%, while employment in plywood and veneer mills declined by 50%.[10] Part of this catastrophic change was due to the severe recession in the early eighties, but there were at least two other contributing factors: mechanization within the industry and a significant decline of the resource base due to causes unrelated to the spotted owl.

Because an increasing proportion of timber is of small, uniform dimensions, automation in the mills has accelerated greatly in the past two decades. According to the estimates of state employment agencies, mechanized handling of materials and computer-aided sawing, peeling, and grading have reduced the number of workers needed to process lumber by an average of 1.2% annually since 1970.[11] According to the Society of American Foresters, it took an average of 10.6 workers to produce one million board feet of lumber in 1976 but only an average of 8.6 workers ten years later.[12] In California, the number of workers it took to process one million board feet declined by 33% between 1960 and 1990.[13]

How do these numbers compare with estimates for the decline of jobs due to spotted owl set-asides? According to two separate studies, the unemployment due to automation will amount to more than twice the unemployment due to the strictest possible preservation of spotted owl habitat over the next twenty years.[14] According to another projection, employment would continue to decline in the Pacific Northwest forest products industry even if the harvest levels actually increased by 20%.[15] And in fact, employment did decline by 14% between 1980 and 1988

despite the fact that the output of finished lumber increased by 19%.[16]
Regardless of environmental constraints, loggers and mill workers will
continue to lose their jobs at a rapid rate.

Large mills owned by national corporations are particularly prone to
replacing workers with machines. When Weyerhaeuser invested $400
million to modernize its mill in Everett, Washington, the number of
employees dropped from 900 to 500.[17] Although large companies
would appear on the surface to offer greater stability than small ones,
this is not in fact the case. According to a study of mill towns in Oregon
and Washington, employment remained more stable when distributed
among several smaller mills than when dominated by one large plant.
Since the smaller companies are less capable of devoting large quantities
of capital to automation, they often try to compensate for their competitive disadvantage by recovering the highest possible return from
each log. This greater attention to detail and maximization of value-
added production translates into more jobs for community residents.
Interestingly enough, the threat that monolithic employers present to
community stability is reflected not only in the mills but in the ownership of the resource base: the greater the percentage of timberland
owned by the largest single party in a given county, the lower the
median family income.[18]

Increasingly, automation threatens to expand from the mills to the
actual harvesting of timber. On many sites, the increased uniformity of
smaller, second-growth trees makes it possible to cut, buck, yard, and
load by machines. It would seem at first glance that the greater amount
of limbs on second-growth timber, combined with the increased
number of stems required to produce a given amount of lumber, would
call for more workers in the woods. In fact, however, the number of loggers required to harvest timber has actually been declining in recent
years as the percentage of second-growth logging has increased. One
study reported a 9% loss of logging jobs per decade for each million
board feet harvested;[19] another study of worker productivity in California between 1980 and 1990 reported an 18% decline in the number
of loggers needed to produce a unit of output.[20] Even in the Pacific
Northwest, where the terrain tends to be more troublesome than in the
South, mechanized harvesting is beginning to play a significant role.
According to one estimate, 60% of the terrain in western Oregon and
85% of the terrain in eastern Oregon is suitable for harvesting by
machines if the timber grown on these sites remains less than 20 inches
in diameter.[21]

While the number of workers required to cut and process a given unit of logs has decreased, the amount of timber available for harvest has also decreased. Historically, the timber industry has cut trees on its own land at a rate which did not even approximate sustainability. The industry did not expect to seriously jeopardize the timber supply, however, for it expected to be able to harvest the vast stands of old growth on government lands as it waited for its own plantations to reach marketable age. The industry reasoned that it had only to buy a little time until the intensive forestry techniques it was applying to its own holdings would begin to pay off with greatly enhanced yields.

During the 1970s, timber interests pushed for—and received—increased cuts in the National Forests. As Forest Plan inventories (mandated by the RPA and NFMA) were tabulated in the 1980s, however, it became obvious that the level of cutting which the Forest Service had been permitting was not sustainable. Years before the spotted owl set-asides, the new Forest Plans reduced the allowable cuts on National Forests in Washington, Oregon, and northern California by an average of 5%.[22]

Supposedly, the timber industry would be growing more timber as the last of the government old-growth was being harvested. In fact, however, cutting on industrial holdings continued to outpace growth; in 1986, for instance, net growth of softwood growing stock was only 77% of the softwood harvest on industrial lands.[23] Why did the industry cut more timber than it grew at the very time when it had planned to be replenishing its stock?

In part, the industry might be the victim of its own unrealistic expectations. Perhaps it underestimated the extent of the damage done by previous high-grading and unsound logging practices; perhaps it also overestimated its ability to turn things around with intensive forestry. Although the new stands of man-made plantations can still turn a profit, they do not really produce that much timber. Since the industry has been cutting its trees at economic maturity rather than productive maturity (see above, Chapter 7), it needs to cut more and more of the smaller trees in order to feed its mills.

And if it can't keep up the cut, the mills have to be closed. As the resource base has diminished in the Northwest, the industry has looked elsewhere in order to stay in business. Between 1978 and 1990, the seven largest timber companies in the nation reduced their manufacturing capacity in the Northwest by 34%, while simultaneously increasing their capacity in the South by 121%.[24]

Machine harvesting: the feller-buncher

With less timber available from public and corporate lands, the industry has tried to convince private, nonindustrial holders to harvest more of their growing stock. As the available supply shrinks, prices naturally go up—and the cutting of trees becomes more and more attractive. In many areas, mill representatives have canvassed landowners vigorously in an attempt to buy stumpage. At least to some extent, they have been successful, for it is hard to resist the quick and easy profit generated by these higher prices. But in the long run, this kind of haphazard, unplanned harvesting only adds to the timber shortage. It constitutes a negative feedback system: as prices continue to climb, the tendency to cut becomes stronger than ever, thus adding to the shortage in the future; this drives future prices even higher, which in turn serves as an additional incentive to hasten the cut, and so on. In the words of D. F. Flora and W. J. McGinnis, with this kind of shortage-driven market "it is hard to imagine a reserve supply of timber lingering on the stump for long."[25]

The demand for timber is stimulated even further by the export market. Japan and other Pacific Rim countries are willing to pay dearly for this resource that they themselves are unable to produce in sufficient quantities—and they pay particularly high prices for timber that has not yet been processed, for they prefer to mill lumber to their own specifications and with their own workers. But the sale of unprocessed timber to foreign countries is of no benefit to the mill workers here in the United States; consequently, the export of whole logs from federal lands has been banned since the 1960s. (Recently, this ban has been extended to include state and local lands as well.)

Although logs-for-export are now unavailable from public land, the private sector is more than willing to sell to the Japanese or any other buyer. In fact, until recently, some companies even received special incentives for the selling of unprocessed logs; by conducting the transactions through specially designed "Foreign Sales Corporations," up to 30% of their profits would not be taxed. Several large corporations, including Weyerhaeuser, the nation's leading log exporter, took advantage of this tax break—much to the distress of mill workers in the Pacific Northwest. As of this writing, Congress is considering a proposal to terminate this incentive. Even so, the export of raw logs will continue, aggravating unemployment in the Northwest.

All these factors—legally mandated limitations on old-growth harvesting, mechanization, previous cutting at unsustainable levels, more realistic Forest Service plans, and the export of unprocessed logs—send a strong but unhappy message to timber-dependent communities: in this day and age, employment cannot continue at historic rates.

Walter Smith

Former Logger

"My whole family comes from a logging background. My dad was born in Coos Bay, Oregon. His father, my grandfather, worked loading logs on ships, and a lot of my uncles worked in the woods. After World War II, when my dad got out of the army, he cut old-growth redwood in Marin County. He was still hand falling, using an ax and a cross-cut with a partner. The early chain saws were pretty heavy. They used the old drag saws for bucking, but most of the fallers still preferred hand saws. He had lots of accidents. He actually got pinned between the seat and the floorboard of an old tractor. A log rolled down on the cat and pinned him because there was no canopy in those days.

"By the 1950s everybody was using chain saws. My dad moved north and became pretty well known as an old-growth redwood faller. We moved up to the Eel River, where they had some pretty big redwoods—up to eighteen feet in diameter. I became part of a community in Holmes where everybody worked with timber. Next door was a choker setter; next door on the other side was a person who worked in mill; then a timber faller, a cat operator—they all lived on the same street.

"As a kid I went out with my father, packing his lunch and carrying his tools around and watching him cut these giant trees. We used to go and enjoy the redwoods. We would go for a picnic to the Rockefeller Forest, where we would just be in awe of the redwoods. Although my father cut them down, we also had respect for the forest. There was kind of a funny two-way thing going on there, which was true of a lot of the people there. I remember him telling me later, after the Forest Practices Act came about and there was this whole new idea that the environment needed protecting, he said, 'I look back and I'm just astounded at the things that we did—that we would run cats down the middle of the streams and cover them up and put landings in the middle of creeks and fall trees right in the middle of the river.' He realized how short-sighted they were.

"But in those days there wasn't a controversy, no second guessing. Everybody was proud of what they were doing. Probably this was typical of the fifties: America was doing great, and we were all moving forward. I do remember once, though, my first contact with an environmentalist, a guy from the Sierra Club. I couldn't have been more than twelve, and we were down on the river bar. This guy came down in a Cadillac wearing a suit and tie and he got stuck. There were three boys; all of our fathers worked in the timber industry. So we go over and we help this guy and we push him out. He was quite pleasant, but then he asked us what our fathers did for a living. I piped up, "My father's a timber faller." And we just got a barrage about how we were cutting the most beautiful objects on earth and blah, blah, blah. We got the whole Sierra Club spiel from this guy in a cadillac with a suit and tie. From that day forward, when I thought of the Sierra Club, I thought of guys with suits and ties who ride Cadillacs. It's probably still the same today.

"Because of the danger of it, and my dad had been hurt so much, my parents would just [as soon] that I didn't get involved in the woods. So they sent me away to college. But I didn't really fit in down in the city. I never learned anything in school that I could really use. I never got a degree, so I didn't even have a piece of paper that enabled me to *pretend* that I had learned anything. So then I came back and started working in the sawmill and pulling green chain. It took me about three months to realize that my future wasn't pulling green chain. So I went up to my dad one day and said to him, 'I don't want to do this. Why don't you teach me how to cut down trees?' He and my mother weren't thrilled because of the dangers of it. In a logging community, that's something we all lived with: people on crutches, people who had lost a limb.

"But he said okay, so I went to work with him. I learned my trade, and I learned it really well. Even now, I have pride because I have a skill that not many people can do: to cut down a big, old-growth redwood. Basically, it's an environmental skill when you get down to it. There are environmental ramifications for being good at what you're doing: being able to drop a redwood where you want it to go and not hurt the rest of the forest. With redwood, you have to be extremely accurate or the trees will break up into toothpicks and be worthless because of their fragility. Later, I started a timber falling school. I developed a whole curriculum that included not only timber falling but also small engine mechanics and forestry and safety and business skills for logging contractors.

"In 1984 I started a logging business with two partners. We tried to work different places, but the logging economy at that point was really bad. Eventually we went to work for Louisiana-Pacific. After a couple of years, I noticed that LP was running out of conifers. I asked them, 'What are you going to do?' Well, in 1986 they bought the Masonite Corporation's timberland, another 95,000 acres. They went in and started cutting all that, but that only lasted until 1989, when they started running out of conifers again. When LP started to harvest tan oaks for chips, that was the last straw for me. It was the bottom of the barrel, scraping every last bit of economic value out of the forest. This was tree mining.

"Then right in the middle of 1989 it came out that LP was moving to Mexico. At the same time, they had just shut down a mill in Potter Valley, and soon after that they shut down their remanufacturing plant in Cloverdale. We were all aware of this. I had gone down to LP several times and talked to them about the problems with the community, that they were running out of trees and people were scared of losing their jobs. They were saying, 'Oh yeah, we're really concerned about the community, too.' So all of a sudden it comes out that they're shutting a mill down here and moving to Mexico simultaneously. I mean, they must take us for idiots. Do they really think people can't see the connection between opening up a remanufacturing plant in Mexico and one shutting one down in Cloverdale?

"I just blew my lid. I showed up at a news conference with some environmentalists and blasted Louisiana-Pacific. Well, of course I hit the newspapers because here I was with my suspenders and striped shirt and the whole works. I'm working for Louisiana-Pacific as a contractor and I'm saying that this was a real slap in the face to the people of Mendocino County. We got called the next day and told that LP was going to throw us off the land. But we were under contract at that time, and there was no way they could break a contract because of what I said at a news conference. Still, it was made very obvious to me that I either shut up or we wouldn't get any more work. Maybe I could still have smoothed things over, but I felt too strongly about what was happening to shut up.

"I became very political. I wrote public comments on THPs. In fact, because of my comments, several things got changed because finally there was an eyewitness to how they were screwing things up. (I had probably driven on 90% of Louisiana-Pacific's back roads; I knew it inside and out.) For instance, every timber harvest plan that I worked on was called 'shelterwood removal.' It didn't matter whether it actually had a shelterwood story or not. We had some areas that hadn't been cut, and they would just take the biggest trees and call it 'shelterwood removal.' Their idea of 'shelterwood' was any tree that cast a shadow. They would cut a tree that was really a part of the understory but had grown really well. So they were cutting the genetically superior trees, as well the old overstory shelterwood. I started pointing that out, and CDF started changing their attitude. That's when the THPs became more complicated. You couldn't just do, 'We're doing shelterwood removal on 300 acres.' It didn't really help matters any, but at least it made things more truthful.

"When we worked for LP, we were cutting everything in sight. We were hauling in what we called 'garbage' for chipping: downed logs, tan oaks, everything. There was no skill to it at all; you just cut down everything you could and hauled it into the landing. With the big timber, it takes some finesse to move a big log; you can't just wrap a choker around it and the cat pulls away with it. You've got to make the log roll out from behind the stump. The tractor driver has to be able to maneuver it down the skid trail and pull something that weighs as much as the tractor—without damaging other trees. Loggers like to have a challenge; they like to have something to think about. That's why timber falling, for me, was interesting: not a day went by that was the same. It's the mix that makes it exciting. They say logging is in people's blood. It's because of the rush—the danger, the challenge. They say that logging has the highest accident rate of all occupations, because you always have to be on your toes. There's times when you have to sweat blood because you know you're in the most extreme, dangerous situations, but instead of not doing it, you figure out a way to do it. That's what makes it all worthwhile.

"But when it went to hauling garbage, it was like going from painting a mural on your house to painting the whole thing white. It took the art out of it. This is something the logging companies want: they turn it into assembly line work. And that's what they want to do to the forest. You automate the forest, you make it into an assembly line, you package it, unitize it. Whatever you have to do to make it the cheapest way.

"The whole thing about doing away with old growth and turning to fifty-year rotations is that you get a tree that you can deal with easily. The trees are not that different in size. People don't have to think about how you're going to move a tree that's six foot in diameter and a tree that's two foot in diameter with the same machine. Now all the trees are sixteen to twenty inches in diameter at the biggest end. You've reduced it to where it's all basically the same situation.

"It's the same going through the mill. Like the mill in Cloverdale, it only took up to thirty-inch logs, and it produced about 250,000 board feet a shift with less than a dozen people. It's an incredible automated system. When I first started working in the woods, the average at our mill was only 90,000, and there was probably thirty of us to produce it. The sawyer was always looking for grade, cutting it and turning it constantly to get grade out of it. The sawyer was someone who was revered as an artist, because this person was going to make it for you by getting the vertical grain heartwood out of it. Now you get these mills where grade is not important anymore. They don't even make anything other than a two-by-four or two-by-six. They just cram them through as fast as they will go. The guy goes to work and just punches a button and the carriage whips right through it.

"Mechanization is one of the biggest reasons why we're losing our jobs. The mills are all computerized. The companies don't like to talk about mechanization and the loss of jobs. They say that it's the environmentalists who are not allowing us to go in and get the timber; they've got all this land tied up. They're blaming it on the owl, but the whole reason we have an owl problem is that they didn't treat the ecosystem responsibly from the beginning.

"The biggest reason for the loss of jobs is the overcutting of the earth, which of course goes along with the exporting of logs. If we were harvesting in balance with the forest, we'd have a hard enough time supplying just what we need here in the United States. The companies are always saying that they plant ten trees for every one that they cut, but those ten trees have no board foot value when they put them in the ground, while they just cut one down that has a thousand board feet. There's no way that they're growing as much quality lumber as they're cutting. Maybe they can sustain a yield of twenty-year-old trees that can be made into chipwood, but that's about it. They certainly aren't sustaining all the complex functions of the forest.

"The first loggers took over a hundred years to cut through Mendocino County. Even though they were clearcutting the old growth and doing the burning, they covered such small amounts of land at a single time that the impact on the overall ecosystem was less. But when we finished with the old growth here in Mendocino, all of a sudden—bang—it took only twenty years to go through again and get all the second growth. Now the watersheds are just beat to death, and the jobs are gone.

"We created this system that just gobbles up enormous amounts of timber in very little time. If you're producing 250,000 board feet a shift and you're running two shifts, that's 500,000 board feet a day. Running 200 days, that's 100,000,000 board feet in a year. It doesn't take long at that rate to go through quite a stretch of ground—especially when we're talking about second growth, which doesn't have the same amount of board feet per acre that we had in the old growth.

"It used to be when the timber disappeared, people could go to Oregon or Idaho or somewhere else to get a job. Now, when you're out of a job here, try to find someplace in the whole Pacific Northwest where you can get a job. Maybe in Alaska, but I even doubt that at this point.

"There was a period of enlightenment in the '70s with the new Forest Practices Act. For awhile there, people really thought the timber industry might be coming around to a more environmental approach. There seemed to be a promise in those years of a sustained yield. It looked like there would be more long-term work in the woods. It seemed that Harwood, where I was working, was actually on a sustained yield.

"But there was a marked change after '82, when we had this timber depression. After that it was just gung ho, from that point forward. The bottom line was the only way to do business. In '85 or '86, timber companies started making loggers bid for contracts rather than just negotiating them on their neighborhood turfs. LP used to have foresters who managed the same watersheds, who knew what the watershed was like. They knew where all the old roads were and all that stuff. The workers and the contractors usually came from somewhere nearby. Then all of a sudden they decided that wasn't good enough. They probably decided the foresters were getting too buddy-buddy with the contractors, so all the foresters were changed around. They had different assignments in different places. They had all these people bidding and going from one place to another, clear around the county.

"It was just a complete change. Trying to make a living at logging became very, very difficult—and to pay your people. My partner and I felt that we couldn't even do the business if we couldn't pay people a decent wage. Back in '78, even most choker setters were making in the $15,000 range; that was quite a bit of money back then. Well, they're not making any more now, sometimes even less.

"Starting in the 1980s, there were a lot of contractors that hired illegal Mexicans to save money so they could keep their bids down and get work. The problem was that if an illegal Mexican got hurt, he didn't want to get deported along with his family, so he didn't want to go to the hospital where people would ask too many questions. A lot of times they would just disappear; you never knew what happened to them. It was almost like in the early part of the century: when a logger broke his leg he was worthless and he got fired. They got somebody else who could do the work. There was no disability or workmen's compensation or anything like that.

A choker setter: will he lose his job?

"One thing that has to be mentioned when we talk about cutting jobs is that we had *too many* jobs. We created this false economy because there was no way we ever could have continued cutting at that level. If we had started by cutting the amount of people in half, we probably still would have that same number of people working today and we would not have the perceived instability in timber communities. But it's hard to tell that to an unemployed logger.

The people of the woods are hurting. Even without the spotted owl, mechanization and the depletion of old-growth timber would make many of their jobs obsolete. Although the timber industry blames the unemployment on environmental set-asides, it is the structure of industry itself that is responsible for most of the lost jobs. And yet the spotted owl has certainly accentuated the problem, hastening the con-traction of timber-based economies. V. Alaric Sample and Dennis C. Le Master, in their thorough analysis of the economic effects of spotted owl habitat protection, concluded: "The need to maintain old-growth forest habitat . . . collapses the time available to local communities to make the necessary adjustments. However, the change is fundamentally driven by factors other than habitat protection. This issue did not precipitate the situation, and allowing species dependent on old-growth to go extinct will not resolve it."[26]

As environmentalists and the timber industry argue about who is to blame, the people of the woods are caught in the middle, left alone to suffer the consequences. What can be done? Is there any way to put the people back to work without further damaging the forests or depleting the resource base?

11

PEOPLE AND THE FOREST ECOSYSTEM
Looking to the Future

It is ironic that unemployment runs rampant at the very time when there is so much work to be done in the woods. In California's Trinity County, for instance, official unemployment is over 20%, but that doesn't even come close to reflecting the magnitude of the problem. Many people whose benefits have expired are no longer on the list, having lost all hope of finding a job. Some families have moved out of the area, while others have kept minimal roots at home as they drift around Northern California in search of work. According to one local resident: "If you sit down at the base of Hayfork Mountain, you can see the mass exodus out of here on Sunday night. They come home for the weekend, then they go back to work. Everybody lives in a trailer. We're turning into turtles, carrying our house on our back."

Meanwhile, much of the forested land in Trinity County is a wreck. Hillsides have eroded, streams have been filled with sediment, regrowth has not lived up to expectations. Worst of all, the fuel buildup has reached dangerous proportions, caused by a combination of past fire suppression and brushy regrowth in the wake of clearcuts. In the past few years, a series of catastrophic fires has wiped out over a hundred thousand acres of Trinity County's forests.

While environmentalists were blaming the logging industry and the Forest Service for the sorry state of the woods, loggers were blaming the environmentalists and the Forest Service for the loss of their jobs. With everyone blaming everyone, the community was tearing apart at the seams. Then, like a flash, environmentalists and loggers suddenly realized their common interests: healthy forests and jobs, far from being contradictory, could go hand-in-hand.

Nadine Bailey, *Industry Spokesperson*

Joseph Bower, *Environmental Activist*

Nadine: "My great-grandparents homesteaded in a place just southeast of here, Indian Valley. I grew up with tales of how you had to ride everywhere on horses. My grandfather had the first car in Hayfork. I grew up with a real sense of history, looking at all the old pictures. The other thing that influenced me was I spent a lot of time with my dad out in the woods. He always wanted a son but he got a daughter, so instead of playing with dolls I went logging. I've always had this love for logging and the people who were involved in that industry.

"When I got out of high school I worked on a brush crew, did some right-of-way falling, actually worked in the woods, which is in the minority for the women around here. I worked enough so that I knew that it wasn't something I wanted to do to make a living. Later, my husband and I ran a logging business. About the most employees we had at one time was fifteen, almost a half million dollars of payroll a year. We were one of the bigger employers in the county."

Joseph: "I don't have a long history in Hayfork like Nadine. When we moved here 20 years ago, we didn't have it in our minds that we would be involved environmentally. But soon after we arrived, we found out that they were spraying 2,4,5-T and 2,4-D and in some cases leftover stocks of Agent Orange on the forest. That got us rather excited, so we started asking the Forest Service about their spraying plans and ran into a bureaucracy that basically said, 'We're the government, get out of our way.'

"That was the beginning of our environmental activism here. It soon became apparent to us that the Forest Service was mining this country rather than practicing sustained yield forestry. For the past 20 years we have been trying to reform that type of management, trying to educate the Forest Service on watershed and fisheries issues and ways to get away from clearcutting."

Nadine: "It was Joseph's fault that I got involved. It seemed like he would never be satisfied until there was not a logger left in Trinity County. I watched him on a TV program one day; he was saying something about 'fuzzy clearcuts' that I didn't feel was quite accurate. I said, 'If he can do that, so can I. The next time they come to Hayfork they're not just going to get Joseph's side of the story.' I spoke at a high school rally, and then the Forestry Association asked me to come speak at a news conference. People liked what I said, and it just started snowballing. Since then I've been active with Women in Timber; right now I do consulting work for the California Forestry Association, which is the industry association for milling and timberland owners.

"I think what I've done is put a voice to what was in the hearts of people who never spoke up for themselves. I talk about what loggers really do, which is provide a product that everybody uses. Loggers take pride in their work. Many people don't understand this; they make it sound like we go out and cut trees and then the trees get sucked up into the air and nobody ever sees them again."

Joseph: "That thing about the clearcuts turned out to be an inaccurate assessment of what I was saying. Nadine found that out as soon as we started talking."

Nadine: "We had been listening to each other's sound bites. But then I went up to the Timber Summit in Portland, and President Clinton, in his closing statement, said something like: 'Go home and sit in the conference room instead of the courtroom. Try to work these things out together.' So when I got home, I stopped Joseph at a Board of Supervisors meeting and we started talking."

Joseph: "We actually had a forum that facilitated our getting together, a bioregional group that covers the Trinity watershed. I suggested to Nadine that she come. It's an open forum where anybody can attend, and we work on consensus. At first I thought it would be impossible to come to consensus, but it's worked amazingly well."

Nadine: "It's like teeth pulling."

Joseph: "Yeah, but it still works. We haven't ever had anybody block consensus. And its a very diverse group. You name the interest and they're there. But dealing with a large group of people, it's hard to get down to the nitty-gritty. So we decided to form a working committee. Nadine and I, along with a couple of other people who represent timber interests and a couple of people who represent environmental interests, started getting together on a regular basis.

"It was easy to agree that there was a lot of work that needed to be done in the woods. We immediately came to consensus on the need to invigorate the local economy. (Since the Dwyer injunction on the spotted owl, it's greatly curtailed activity in the woods around here.) We were also in agreement that we wanted to restore the anadromous fishery and that we wanted to fireproof and flood-proof our forest and watersheds. We sat down and wrote a proposal to President Clinton that said: 'Here's the things we would like to do, and here's how much money it would cost. We're in agreement in our community. If you send us money, we will do these things that will help restore the health of the forest and provide full employment for anybody that wants to work in this county for many years.'"

Nadine: "This is the first time anybody has gotten the two sides together. The whole problem has been this knee-jerk reaction. Our only hope is to not have these emotional rock-throwing contests. We have to sit down and work out solutions to our problems. That's where Joseph and I agree one hundred percent. We're sick of whiners who don't do anything. These complainers who just want to say nasty things, when you sit down to work they disappear. This is very difficult work, very time consuming. You find out who really cares about the resource, and who really cares about the people."

Joseph: "There're two parts to the plan we came up with. The first is forest and resource rehabilitation; the second is for a unique educational facility in Hayfork. It would be a training and research center. It would not only train local people to work in the woods, but also train agency personnel and private foresters on the New Forestry techniques.

"The first part has four components to it. One is just to get enough money to the Forest Service to maintain the roads they already have. You see how steep the country is, and when you don't maintain your roads they sort of melt downhill into the streams and ruin the fish habitat. The second component is to do road rehabilitation. A lot of these roads are built on landslides and they've slipped out. They need to be closed, with their drainages pulled so there's no more culverts to plug up.

"The third and largest component is fuel reduction. We're about ninety years into putting out fires in this country, and during that time the ecology has changed greatly. Here in Trinity, before we started putting the fires out, the country burned every seven to ten years. That created a forest where there wasn't any accumulation of fuels and fuel ladders. There are people around here who are eighty years old, and they say when they were young you used to be able to get on your horse and ride right through the forest, anywhere you wanted to go. Now, you can't ride a horse off the road, the forest is so thick with brush and undergrowth. That's why we've had these catastrophic fires. In 1987, 60,000 acres burned up just on this one ranger district.

"We don't want to lose what we have left to fire, so the obvious thing to do is to restore more natural conditions by taking out this excess fuel. In the process, we will create a tremendous amount of employment and bring out a tremendous amount of wood, whether it's a traditional saw log or something else. We're advocating doing that on a pretty ambitious scale, like 25,000 acres a year.

"The fourth component of the proposal is to keep up the projects that are trying to restore the anadromous fishery to the Trinity River system. There are fourteen different agencies working on this. Although there have been some problems with the programs, we're asking that we continue for another five years so we can get the fish back because that can be a tremendous economic asset. Once upon a time, people came here from all over the world for the salmon and steelhead fishing.

"We're only asking for $190 million over ten years to do that. Although that sounds like a lot of money, the federal government has historically taken more than $100 million a year from Trinity's resources for the federal treasury."

Nadine: "Talk about mining, the government has mined Trinity County for money from the beginning. It's been a positive cash flow straight into the federal coffers. All we're doing is saying: 'Wait a minute. Some of that money has got to stay in the county to do some of the things that have to be done.' Whatever the government decides—whether to log it all or not log anything—they still own forest land that they have to take care of. It costs a certain amount of money to own timberland, and you have to pay people to do that. You cannot be an absentee landowner any longer. (That's Al Gore's little quote.) You have to fund projects like ours. It's really a simple project. It says, 'We're going to take care of the forest. There's things that have to be done—and here's the bill.'"

Joseph: "They're telling us now that there is no new money. It's just going to be reprogrammed money from within the Forest Service budget."

Nadine: "They're going to take it from other districts. Already people are mad at us for even trying to think up this Adaptive Management Area, and now they'll be even more angry for taking money out of their districts. [The Hayfork/Trinity plan as been accepted as one of the ten projects to receive special attention in Clinton's Forest Plan.] It's going to create some real problems. And while everyone debates these problems in the political realm, the forest will continue to create a fuel buildup. See those thunderclouds over there? It happens every August. If we don't act soon, the forest we've been fighting over might disappear before our very eyes."

Joseph: "The public expectation here is that something is going to happen in the woods next year. The Forest Service is saying: 'That's too soon. We can't do anything by next year.'"

Nadine: "They'd better get something going, because we may both have to leave town if this doesn't work."

Joseph: "I see it as a real problem if it all fails. Two years ago, this community was on the edge. People were threatening to lynch me and burn my house; they were feeling pretty desperate. At the present time, people are feeling optimistic and hopeful that things will happen, but if they don't, things are going to get nasty again."

Nadine: "I went downtown the other day, and there was this little girl (she couldn't be more than three or four) in her nightgown, hair not combed, sitting on the steps of the bar. That's the weight we have sitting on our shoulders. If this doesn't work, it will be the kids of this community that hurt. There will be no future. It will turn into the new Appalachia."

Joseph: "But it's also a marvelous opportunity. These are things that we've been talking about for a decade, and now there's a potential to see it happen."

Nadine: "I hope all this isn't just political rhetoric, like so many things are. I think I'll let Joseph tell them if it doesn't work out.

———— ” ————

Even if the Hayfork/Trinity Adaptive Management Area receives full funding, will former loggers be able to "adapt" to thinning brush and repairing streams instead of cutting timber? Historically, loggers have placed a high premium on getting the logs to the mill; occupations like tree planting and brush clearing, on the other hand, have had a relatively low status, being the province of "hippies, wetbacks, and bums" (in the words of a veteran catskinner). Restoration work might provide employment, but the people of the woods will still have to overcome cultural prejudices if they are to adjust to the new realities.

Bob Ford

Logger, Mill Worker, Restoration Worker

"I was born in Willits, 1931. We moved around a lot. My father made split stuff from old-growth redwood. He did tanbark work, too. They had some big old-growth tanbark, three or four feet diameter, maybe even five. We'd peel the bark green, an ax handle and a half, somewhere around five feet in length. We'd take one ring off and fall the tree, then peel it on the ground. We'd take the bark and just leave the wood to rot. We hauled it out to the road with mules, Humboldt pack saddles with hooks, piled it right over the mule's back. I worked the mules, I liked that—and then when I got older I worked the truck. I liked that even better than the mules.

"My dad paid me too much. The idea was that I'd take the money and go to college, which I did. From then on, I worked on my own. One summer I worked for tree surgeons. That fit right in with what I had been doing, working with chain saws. We had one of the first chain saws in the woods, the old handle-bar type. And I could climb trees, because I didn't have any problem with heights. I had already taken flying lessons. I soloed in a plane when I was sixteen years old, and I had my pilot's license for my 17th birthday.

"The next summer I worked as a choker setter. We were working that old-growth timber on steep ground. It was pretty dangerous that year. We had to pack the boss out with a broken hip. His cat run away. Then my partner got run over with a TD-24, killed him of course.

"When I got out of school, I went in the Air Force. When I finished with that, I went back to work in the woods. I got hurt a couple of times. Tree hit me in one knee, and a logging trailer hit me in the other knee. I should've went to the doctor and took care of it, but I was too macho. Went right back to work both times. After that, I go to put the brakes on in the cat and my leg would lock. Not very safe. My knees started giving out before I was thirty years old; just walking along, and I'd start to fall. Years later, a doctor took a look at my knees and thought I'd been in a car wreck. I said, 'No, just worked in the woods.'

"So I went to working in a sawmill in Fort Bragg, actually a planing mill. I worked there for about ten years, but I had a flying career going too. I did charters and managed the Little River airport, and then I went to crop-duster school. I did crop dusting in Corvallis, Oregon. On Christmas tree farms, we were putting on stuff to kill the grass; on tree farms, we were putting on stuff to kill the broadleaf weeds. I worked for timber companies up there, flying corporate bigwigs around in British Columbia. Over in Idaho, I flew air patrol for BLM, the Forest Service, and about everybody else. In Utah and Nevada, I flew cattle buyers around. On them ranches, they didn't have separate landing strips; they just landed on the roads.

"Then I came back to Fort Bragg where my wife's parents lived. Walked in and got my old job back. I could see things were changing. Earlier, you had all nice clear redwood, vertical, nice tight grains. Now, all you had was sap. And when you did have a clear piece of wood, there was a long ways between grains.

"When my wife and I broke up, I left my job and went up to Redding. When I came back, I couldn't get a job. Went on disability. Finally I got a government training job, learning a lot of things about forestry and restoration and revegetation that I never knew before. I had worked in logging before, but I had only learned how to take things out, not put things back. I had never figured on the long term. I always figured after we logged it, that would be the end. A lot of times, after choking on dust all day, all I could think about was where I was gonna get my first beer.

"I've thought about how we did stuff in the old days, and how they do stuff nowadays. They're a lot more concerned for biological things than we ever were. We were just concerned about getting the logs; we didn't worry too much about what we did to the environment in the process of getting them. I guess, had we owned the land, we'd have been more concerned about it. But that's not how it worked; we owned the timber, not the land. Even back then, I noticed that the landowners were more concerned about the environment. My grandfather, he owned his land and he thought ahead a little bit; he didn't just tear everything up. He used horses for his logging until later, and then he used a little D-2 or D-4 or something—he didn't go in there with a D-8.

"Even back in the fifties, we knew we weren't supposed to do things like drag logs right through a blue-line stream. I saw it happen; actually, I did it once myself. I knew that if I got caught I was in trouble, but it was the quickest way. Then I came back the next day and I saw dead fish laying there—not really dead yet, just slopping around, couldn't breathe from all the mud stirred up. I love to fish, so I could see right there that this wasn't right. But I didn't give it much thought until I got involved in this training crew.

"They paid us minimum wage. My unemployment had run out, and I was glad to get anything I could. We did some splitting, and I had some experience in that from when I was a kid. We did trail-building and stuff like that. It was good outdoor work, which is what I liked in the first place. It wasn't as much heavy equipment involved like in the fifties and sixties, so you wouldn't get hurt so much. It was more like it was in the first place, when I went into the woods with my dad and made split stuff and peeled tanbark. I like that.

"But in this restoration work, there aren't very many people like myself who grew up in the woods with logging. Most of them are younger people who don't seem to be natives of Humboldt County. When I trained, some of what they learned was just things like how to camp out in the woods. These were things which I already knew. The trainers, they said that they actually learned stuff from me—mostly mechanical skills, like taking care of chain saws or any kind of equipment.

"When I finished with the training, me and two other people formed a restoration company. We called ourselves PFM, Progressive Forest Management. We held out for three or four years. We planted a lot of trees and opened up a lot of brush so the land could be planted. We did stream restoration, too. Now we've joined up with RCAA [Redwood Community Action Agency], which is a bigger outfit. More security that way.

"Getting the old 'Humboldt crossings' and straightening them out, that's been one of our main concerns. Get some fish habitat. In the old days, much too often, we come to a stream and instead of putting in a culvert to save the road, we just dropped a bunch of logs or debris or whatever into the stream and bulldozed dirt right across the top of it. Very simple, and as far as we were concerned, very effective—although it wasn't effective in the long run, because it ruined the fish habitat. So what we have to do now is go in there and get rid of these things and stabilize the streambanks and so on.

"This environmental work, it's very satisfying. You go back and see something that was practically a moonscape, very little vegetation, and now you look at it ten years later and it's all forested. More water than you ever thought there would be, nice clear water. I don't make a whole lot of money on it, but it's a living. And I get the exercise. If I didn't get this exercise, you see I'd be a little over-weight.

"The problem is, there's a stigma about this kind of work. People associate it with long hair, hippies, and like that. And you're working with a lot of women. People think it's not macho enough because there's women doing it too. And we're not dealing with big trees anymore. All we got is these little bitty trees we're putting in the ground. But those little two-foot trees get big; people never think of that part. And you go out there and plant a thousand of those things and you'll feel it, I guarantee you. I think it's a lot harder work, personally. It's a lot easier to walk up to a big giant tree and fall it than to go out there with a hoedad and plant several hundred trees—and do it right. Physically, it's more demanding, but the image is different. You don't go out there with the biggest D-8 they make and pull a big winch line with a choker on there—that's macho stuff.

"But from what I've observed, there's a stigma the other way too. If some logger walked into the office tomorrow wearing his cork boots and applying for work with some restoration outfit, I don't think he'd get hired. These restoration people, they'd think: 'This sucker, he's nuts. He'll cut down all the trees.' They wouldn't trust him. They'd much rather hire a student in biology. So that stigma goes both ways.

"But if they could only get together and understand each other, they'd probably learn that they both want the same things. For one thing, they both want to work. And as far as the forests are concerned, they both want it to be there in the future so their kids can get a job. And for that to happen, they need a healthy environment. They have a lot of things in common if they could just understand each other.

"Personally, I'm kind of a cross-over guy; I see it both ways. But on the other hand, I'm kind of an orphan too, because I don't belong to either side. So I feel a little lonely sometimes. But I do feel good about the work I do. You get to like that green stuff. The trees and the money, they're both green.

———— 〞 ————

At least on the physical plane, loggers are strong on adaptability. To cope with the infinite variability of circumstances they face on a daily

New jobs, new challenges

basis, they have learned to "do what it takes to get the job done." Now, however, their ability to adapt is being tested in a different way: Will they be able to maintain their independence and their rural lifestyles in the face of the catastrophic changes that are occurring in their occupational field? Can they adapt socially and economically to the new tasks at hand?

They probably can—if they are given half a chance. The problem, of course, is to keep that "green stuff" flowing. The need is obviously there: the woods must be repaired, or at least tended more carefully. But who will pay them to do the job?

The first answer that comes to mind is: "the government." But who is that? And through what mechanism? Throughout the Northwest, economically impoverished timber communities are trying to put together environmental repair programs like that of Hayfork/Trinity. Often, these programs are being created through the joint efforts of former enemies: loggers and environmentalists. These new coalitions of local residents are petitioning governmental agencies for funding; in many cases, they also look to private foundations. But the job is simply too big. Competition among communities for scarce discretionary funds does not address the sweeping issues: full employment and environmental health on a regional scale.

In Plumas County, California, the Plumas Corporation has been unusually successful in obtaining funds and implementing projects. But the effect on the local economy has been negligible. This has caused some of the local organizers to wonder: Can a sustainable future be built upon "something so ephemeral as a grant here and a grant there"?

Leah Wills

Erosion Control Coordinator, Plumas Corporation

"My background is anthropology and economics. I've always been interested in the relationship between culture and the landscape. I come from the midwest, and the farm crisis there had a big impact on me—watching a whole way of life collapse because of the collapse of a federal support system. I was surrounded by people who were very affected, so it got me thinking about the relationship between sustainable cultures and sustainable economics, even though I didn't have words for that at the time. In college I looked at two cultures, Native Americans and the Japanese around World War II, trying to understand how they dealt with rapid culture change.

"I left school for awhile and went to work in the central Sierras, doing your basic handwork on trails and so on. I was very impressed with the idea that here were these really rich landscapes, yet the communities were impoverished. I couldn't put it together, so I decided to go back to graduate school. I studied rural development, where I was able to pull in geography, planning, and economics. Then I went to work for the state in community development. Out of my schooling, I had this notion that if you could build an infrastructure—roads, schools, housing—you will attract capital. The people will come and the economies will flourish.

"So I went to work in northeast California, putting in water and sewer and housing. To my complete horror, the economy never did much of anything. We had all these structures and people did move into them, but nothing spun off. The whole concept, which was developed on urban models, didn't translate to rural communities. I felt really betrayed by my education.

"At that point I decided to do my thesis on the community here in Plumas. I focused on resource dependency, federal ownership, and how these affected community stability and community economics. I went back to work for the Forest Service on a survey crew, and my job was to go up and down land lines. We had the old notes, which described the landscape back in the early 1900s. We were running the same lines, but it wasn't the same landscape, not even close. The pieces of the puzzle were starting to fit together: the conversion of landscape, the ownership, community instability.

"When I started to put these pieces into a model, it looked more like the Third World than the developed world. Resource communities like this seemed very susceptible to 'colonialism,' for lack of a better word. They really had to struggle against being simply an extractive, raw resource community. To change this, they would have to turn dollars over and over and try to keep more command of the process, rather than just extracting the raw resource. That would involve some kind of political action, as well as understanding the ecosystem, as well as trying to attract capital.

"At about that time PG&E was talking about spending a million dollars a year to dredge their reservoirs in the Feather River Canyon. I thought: here's all this money coming into the community, but very little that will spin off. A job of that magnitude would probably go to an outside firm. And I asked myself: Where's all this silt coming from? Where's the erosion that's creating it? It was a radical idea, but PG&E actually started looking at whether intervention upstream could reduce their dredging costs. They were willing to pick up 20% of the costs for some projects upstream, since about 20% of the silt eventually became trapped behind their dams.

"We started a demo project while we were contemplating the big picture. People just have to see action on the ground or all the energy falls apart after awhile. That's been a cornerpiece of our process ever since: we start demonstrations in every area, knowing that you might just be treating the symptoms, but at least you give people the idea that something can be done.

"Over the past five years we've done about two and a half million dollars of restoration work with about twenty different grantors or investors. We're real proud of our program. We're monitoring everything, and we can see that lots of wildlife and fish are coming back. But we're not seeing a large impact on economic stability. Two and a half million dollars may sound like a lot, but all it amounts to is about 1% of the economy of Plumas County annually.

"What we really need is a restructuring to a sustainable economy, which hasn't happened because the grant situation is so sporadic. On the environmental end, you put something in and it recovers, and if you develop a realistic long-term management plan, you're okay for that area. You pick a unit and you fix it. But with the economy it's different. People can't restructure around something so ephemeral as a grant here and a grant there. What we're missing is a way for capitalism to move on to the next step, which is to treat ecosystems as capital. Capitalism has always treated resources as a free good, and now we're not in a position to do that anymore. We're obviously working with limited resources, which is not a concept capitalism has worried about in the past. But if you treat an ecosystem as capital, you get sustainable outputs which you can live on, which is your interest.

"But for your capital to be sustainable, you have to talk about capital maintenance or reinvestment. What kind of reinvestment patterns do you have? I see four basic types:

"[1] This particular watershed is in the category of hydro and power and forest products. In this category you can put everything from the Catskills to the west slope of the Colorado to the Pacific Northwest to the northern Sierra Nevada. The resource consumers are the power consumers, the water consumers, and the Forest Service, which in this case owns 84% of our watershed. The number one recipient of money from the forest products is the United States Treasury. Every year, 75% of the receipts get sent back to Washington. When I went back to Washington this year, a rancher here wanted me to ask the government people this question: 'If you're a Third World country and everyone's blaming you for exploiting the resources, but you're shipping 75% of the profits back to the parent company, which is not reinvesting in the local resource base, then what do you call that?' The people in Washington had to squirm, because it was obvious that this describes the situation here. You cannot blame the rural communities for this sort of thing. They've been through a lot of boom–bust cycles relating to policy. I don't look at a community like this as timber dependent; I look at it as federal policy dependent, which ties us in with the rice growers and the farm crisis in the Midwest and to any community which has coevolved with federal policy.

"If the users of power and water and forest products involve themselves in restoration work, they are reinvesting in the resource base. They're also doing the groundwork for the ecosystem monitoring which is necessary for sustainable outputs. At the same time, they're maintaining the kind of infrastructure in terms of leadership and educated workers who will provide the transition to a sustainable economy. If you don't do that, you'll see all those people leave, a brain drain. And once they've left, the mills are shut down, the operators are gone, all the scientists and well-educated people are gone, you're looking at a social welfare solution. All that's left are the people who are marooned there. And that's an endless 'solution': you'll be paying for it for a long, long time. Without some kind of restoration work, some form of reinvestment strategy, you still have problems with the land as well. Then you're into a corporate extractive economy, and you can go anywhere in the Third World and see that that's not a pretty picture in terms of land stewardship.

"The timing of reinvestment is critical. If you intervene immediately, you have a chance to save the economic and cultural infrastructure of the community. You need to intervene within the first three years of collapse—whether a market collapse or a federal policy collapse—or you've started this cycle which is basically irreversible.

"[2] The second kind of reinvestment pattern occurs in what I call the gateway forests, which lead into the big tourist centers like the National Parks or the pricey wilderness recreation areas like the Adirondacks, the Smokies, the Yellowstone–Teton complex, the Sierras, the Redwoods. For those kinds of areas, the reinvestment strategy might involve taxes on recreational equipment or transitory occupancy taxes, bed taxes.

"[3] The third tier could be alternative product generation, value-added products and marketing traditional products that are sustainable. At this stage, you have to find what the market will bear. People are willing to pay more, there's no doubt in my mind. But are people willing to pay 10% more for sustainable products? 20% more? 30% more? This is where you do your certification. But you have to be very careful not to certify alternative products until you've done the thinking about what's really sustainable.

"[4] Once you've figured out what the market will bear, you can turn that around and surcharge everything that isn't sustainable, whether it's from another country or from this country. This is the fourth tier, which involves everybody. The surcharge is on all resource outputs across the board, where the money goes into a fund for restoration, the development of alternative products, ecosystem landscape development, and monitoring for sustainability. The re-investment is shared by all the end users.

"Once you're at this tier, you wouldn't have so much trouble with free trade. In terms of the Third World, it would be an import tax which would force them to certify their products. The floor would be sustainability, instead of who has the least power in a colonial system. When the pressure is always on improving profits for the parent company, you're going to see wages cut, you're going to see benefits cut, you're going to see mechanization as a way to get rid of labor, you're going to see lower environmental standards and health and safety standards. When you're competing worldwide, what you're competing against is the most impoverished communities on earth. Why does America want to be competing on that level?

"The way you get around this is by understanding what are sustainable products, being able to certify them as sustainable, seeing what the market will pay for those products, and flipping that on its head to put a surcharge on everything that can't be certified as sustainable. This surcharge becomes capital which can be reinvested into the ecosystem. At this point, there's not the political will to get to this stage. I'm talking about a time frame: how do we evolve to this level? That's why we're pioneering in the first three tiers, developing the information we need about the market and about sustainability. Say we find out the people will pay $20 more; therein lies the basis for justifying the tax on what's not sustainable.

"A consumption tax on nonsustainable products would provide the market incentive for sustainability. If the price starts to rise, you're going to generate a lot of interest in alternative products on the demand side: conservation, other ways to do it instead of using scarce resources. A lot of things would trigger automatically through the market. Being an economist, I want it to all tie in to the market. Then it doesn't need so much tinkering with all this policy intervention and all this fooling around. Policy changes are so faddish that they lead to real crisis for rural communities. I want ecosystem management to be the last fad. It *has* to last, but it won't last without reinvestment patterns that are built right into the structure of the market.

"It's kind of a wild idea, but I don't see any other way to do it. The taxpayers cannot continue to pay for all these externalities with the market not doing it. The public does not feel it's making enough money to use it's discretionary income to float bond issues to clean up the environment. It's not happening. The taxpayer is hurting right now. If we make that shift to seeing the ecosystem as capital, the money has to come in on a different level, as part of the cost of doing business. The consumer will pay, not the taxpayer. It may end up being the same person, but you have a whole bunch of different forces at work as soon as the consumer starts paying. The market will start telling people what's going on instead of the regulatory system. The market will start kicking in; it will become profitable to pay attention to sustainability. I am an economist, and I think the market is real capable of doing a lot of the work that we've been laying on special interest groups and regulators and legislators.

———— ” ————

Paying the Piper: Who Will Pick Up the Tab?

Here in deficit-ridden America, it is unreasonable to expect that the government will continue to grant sufficient quantities of discretionary funds to repair all the landscapes and employ all the people. Timber-dependent communities in the Northwest do not need a bailout. In the words of forest economist Roger Sedjo, "If society intervenes to save a sawmill in Woodville, isn't it also obligated to save an auto plant in Detroit? An oil refinery in Houston? A farm in Iowa? Tobacco in North Carolina? Where are the limits?"[1] The federal government cannot afford to subsidize every mill, mine, farm, or factory which can no longer make it on its own.

But the government should recognize that the *extracting* of timber does not reflect the true cost of *producing* timber. Historically, the price of lumber, plywood, or paper products has been determined by the availability of timber and the costs of harvesting and processing. But what about the maintenance of the resource base? Shouldn't the price

be determined by the full cost of keeping the forests healthy and productive? Can't we use some of the money generated by timber harvesting to enable the practice of sustainable forestry?

The Forest Service's K–V funding represents a primitive attempt to tie
the costs of reforestation to the receipts gained by harvesting timber.
What we need, however, is some mechanism by which the receipts
gained from timber sales can generate enough money to mitigate and
repair *all* the negative impacts caused by harvesting activities: destruction of fisheries, erosion due to road construction, damage to wildlife
habitat—and, of course, the loss of soil fertility that could diminish
future harvests. Somehow, these funds must be generated not only for
public lands, but for private and corporate lands as well.

An across-the-board users' tax, such as that suggested by Leah Wills,
would certainly do the job. Just as the huge costs of building and maintaining highways are paid by highway consumers via a tax on gasoline,
so too should the costs of maintaining healthy and productive forests be
borne by forest consumers. Who are these "forest consumers"? Some of
us visit the woods a few times a year to go hunting or camping, but we
all make use of the trees every single day as we read our newspapers and
seek shelter in wooden houses. Since the most frequent use of the
forests is the consumption of paper products and building materials, a
tax on these items would be an appropriate way of funding sustainable
forestry.

The problems of implementing such a tax are immense. Is it politically
viable? How would the funds be distributed? And how could such a tax
be applied retroactively to pay for the huge backlog of damage that has
already been done? There are no easy answers, but these are the questions we should be asking.

Realistically, although a users' tax would represent the most comprehensive source of funding, its acceptance and implementation lie in the
distant future, if at all. In the meantime, are there lesser measures we
might take to fund projects that simultaneously lead to healthier forests
and better employment opportunities for the people of the woods?

In Oregon, a ballot initiative proposes to double the severance tax on
timber for large landowners, with the revenues to be used for restoring
ecosystems and helping timber-dependent communities. This would
naturally drive up the price of forest products, bringing them more in
line with the real costs of maintaining the resource base. But it would do
so only for a small portion of timber produced in Oregon. A few manu-
facturers would be placed at a competitive disadvantage, while it would
have no affect on the bulk of Oregon's forests—let alone the forests of

Planting by hand: tomorrow's timber

Washington, California, or Canada. The problem with a severance tax is its selective application. Only by making a severance tax nationwide, and by coupling it with an import tax, would the burden be shared fairly among marketplace consumers.

Another possible source of support for restoration is the certification of products that come from sustainable enterprises. Since certified products fetch a higher price, more money can be pumped back into the forests. Certification serves as a sort of voluntary tax, paid by those who recognize the need to maintain healthy ecosystems. Timber owners who practice sustainable forestry could also receive tax breaks, thereby giving official recognition—and, indirectly, financial assistance—for responsible management. Currently, owners can get limited tax credits for reforestation; these credits could be increased, and their scope could be extended to include restoration projects. As of this writing, there are several bills and initiatives that are placing these sorts of ideas before the voters.

The idea behind many of these schemes is to encourage sustainable forestry by providing incentives rather than by enforcing restrictive regulations. Incentives differ from outright government grants in two respects. First, the word "incentive" is compatible with the ideology of capitalism and in tune with contemporary politics. Incentives are not perceived as government handouts; instead, they are perceived as promoting self-reliance. Second, tax-based incentives, unlike grants, do not really *take* money from the government. Once taxes are placed in the government's hands, they become available for all sorts of worthy uses. Special interest lobbies, many from causes just as admirable as ecosystem repair, compete for any discretionary funds, and the guardians of the public purse are very picky about which projects to support. But with incentives, the government never gets its hands on the money in the first place; it therefore doesn't have to give it up begrudgingly. A ten-billion-dollar tax incentive program is politically viable; a ten-billion-dollar government giveaway is not.

Incentives serve as useful instruments for implementing policy in the private sector. In the public sector, policy can be mandated by more direct routes. Perhaps the United States Treasury should not be allowed to turn a profit with business-as-usual timber extraction until the costs of previous abuse have been paid; perhaps all the money generated from timber sales should be returned to ecosystem management, at least until there is some indication that the damage has been repaired and that the Forests Plans are sustainable in practice, not just on paper. We're talking here about decades at least, perhaps centuries. We need more than computer modeling to prove we're on the right track.

But with harvesting levels now so low, will the receipts cover the costs of restoration? Probably not, at least for awhile—and they certainly won't cover the costs on environmentally sensitive sites that historically have lost money. The distribution of funds should not be tied too closely to the source of the revenues. We should remember that Forest Service K–V funding, although conceptually appropriate, has served as an incentive to harvest timber simply for the purpose of budget maximization. Similarly, the practice of returning 25% to 50% of timber receipts from federal lands to local governments has served as an incentive for local residents to push for increased cuts rather than sustainable forests.

To be fair, any method of funding specific projects should be tied not to harvest receipts but to the needs of the land. But how can this be measured? And how can reparations be made for damage that has been done in the past? Again, there are no easy answers. The people of the United States—as well as the people of Japan and other wood-consuming countries—owe a debt to the forests that have supplied them with undervalued products. This debt will not be paid overnight, if ever—but it certainly should not be increased. We need structural changes in the pricing and taxing of forest products that will insure that no further debts are incurred.

In both the public and private sectors, attention should be paid to generating money by harvesting the by-products of ecosystem repair. Commercial thinning, for instance, often serves several purposes simultaneously: fire danger is reduced, growth is improved for the remaining stand, while hardwoods and small conifers are turned into pulp. But are the harvested trees used to their highest potential? Many hardwoods could be turned into high-grade lumber for furniture making. In older stands, suppressed conifers can have narrow, tight-grained wood suitable for peeling. With the decrease in availability of traditional old-growth timber, tight-grained lumber is increasingly valuable—and it could therefore help pay the expense of fuel-hazard reduction projects.

People in the Woods: The Swiss System Shows the Way

The biggest problem with financing any form of restoration work is that it tends to be labor intensive. Planting, thinning, fuel-hazard reduction, streambank stabilization—these activities are not easily mechanized. They require real human beings who perform hard, physical work and make important, site-specific decisions. But isn't that the point? Don't we want to put people back on the job?

Like it or not, humans are now an important factor in forest ecosystems. We are bound to have some impact on the environment, particularly where we have already established a presence by the cutting of trees. But what kind of an impact will that be? And which humans should be calling the shots? Presently, most of the significant decisions are being made in faraway offices, by businessmen balancing their accounts and regulators passing laws. Wouldn't it be more appropriate for decisions to be made by workers on the ground—the people of the woods—who are familiar with the idiosyncratic needs of each individual site?

Dr. Rudolf Becking

Professor Emeritus, Humboldt State University

"Historically speaking, the European forests were ravaged in the sixteenth, seventeenth, and eighteenth centuries. Wood was the only fuel source available for growing industry and technology. The forest resource was utilized for everything from steam generation to homes and shelter. As a result, the forests were mismanaged and depleted.

"In about 1720, a group of German professors at a university came together to create a new science, an integrated science, to deal with trees and land, to make land more productive and to grow trees. The foundation was laid at that time for modern forestry. Between about 1730 and 1750, they started planting spruce on deforested lands. An agricultural type of forestry was practiced with these monocultures. These stands grew good and dense, and they were harvested in about eighty or ninety years. That was the first generation.

"After the first generation came along, the biological processes changed a little bit. When they started to plant the second generation of spruce in the mid-eighteenth century, the environment was susceptible to parasites and diseases of the spruce. The spruce didn't do so well; the balances were upset by this monoculture. Most of these stands became so depleted by parasites and disease that they had to be cut. The explanation was, 'Our technology wasn't advanced enough. If we knew a little bit better about changing the pH of the soil, about selecting better strains of seeds or growing the seedlings better, or putting on some more fertilizer, this wouldn't have happened.' So the stands were planted again with more professionalism and care.

"Actually, the spruce had collapsed because the environment in the lowlands was outside their natural range. There were a few spruces here and there, but it wasn't right for a monoculture of spruce. So by the 1920s, most of those stands had failed again. That can be documented quite well.

"The third generation of spruce was planted right after World War I. The science was then sufficiently perfected, in terms of the technology and the training of professional people. They planted according to strict agricultural techniques. Still, the third generation was not in its natural place, so the spruce failed yet again, this time more quickly. It lasted only about twenty or thirty years. After

World War II, that generation of foresters who had tried the hardest came to the conclusion that this was not natural. We had to go back to the natural processes. So European foresters have come full circle in forestry.

"The reason that America doesn't understand this is that most early American foresters were sent to Europe around the 1880s to pick up their forestry. They brought back the idea of even-age management, which we still have in this country. But after World War II, when America became a world power, they did not follow the historical development, the scientific development, in Europe. So at present our forestry does not understand what is happening abroad.

"I received my training in Europe just when some of the professional foresters there were moving away from this idea of even-age management. There's not too much forest in the Netherlands [Dr. Becking's homeland], so 90% of our forestry students had to go abroad to get certificates of professional experience. I chose Switzerland, and that's when I was introduced to the Couvet forest. I became intrigued with the philosophy they had there.

"Their basic theory is that the forest is a community of trees. Experience taught the Swiss many centuries ago that when you start to do heavy cutting, clearcutting, then avalanches will develop, and the whole safety of the valley will be threatened because of these steep mountain slopes. It is of paramount importance to keep trees on the slopes, because that is their protection. They've learned and respected that for a long time. So a primitive management system developed, where they only cut a few trees here or there. Experienced farmers were elected by their community to designate which trees would be cut this year and which trees would be cut next year, and so on. These farmers had no formal university training, but they were in tune with the environment.

"Thanks to two foresters, Adolphe Gurnaud in the nineteenth century and Henri Biolley in the early twentieth century, this primitive system was turned into a science. They carefully measured the rate of growth and the changes in the forest over time. They gave this system its scientific foundations. As a forestry student, I became very interested in the scientific foundations of all-age management.

"In all-age management, you thin young stands out, harvest a few big trees, and create a space for the young stands to grow. You work in all size classes. You cut some noncommercial timber; you cut some commercial timber. The advantage of this is that you can more readily balance your books. You get a repeated income. You get small pocket money every week, not just one lump sum at the end.

"The whole philosophy is different. When you have this small pocket money coming to you, you try to utilize these products better. You don't squander them, because you don't have twenty veneer logs—you only have one. And you're not getting another one until next week. So you are more careful. Artificially created scarcity ensures that you don't waste. You make mine props, pole timber, fence posts, bean stakes, whatever you can get out of it. All the wood gets used, even down to the bakeries that go out and gather the dead wood for fuel.

"The other thing—and this is actually the biggest point—is that under even-age management we have always relied on computerized yield tables. We rely on the table to tell us how the stand should grow, but we have lost touch with the stand itself. In all-age management there is no such thing. You have to establish empirically what changes you can expect from the forest, to see how far you can push the productivity of the forest before it stabilizes on your sustained-yield cut. In effect, you have to make your own yield table for every compartment in the forest. You have to learn the limits of the natural system, and you try to maintain that system. You refrain from capitalizing, amortizing, your growing stock.

"The forester is concerned not only with how he's going to get his timber to the mill but with what he's going to do to plant it back. The forester is involved in all stages of succession at the same time, as are the worker, the tractor driver, and the logger. Everybody is directly involved. There is no pecking order. There is no status symbol for the logger who takes out the biggest logs.

"In fact, it creates a stable labor force. The same farmers always come back. There are no contracts where they come in to make this amount of money and then disappear. Some of these people always work the same tract. When the forester gets to a certain tract, he hires a certain guy to cut it, because that guy has been doing it since he was a small boy. A tradition has developed. All the way down to the worker, you have a commitment to a specific piece of land. And specific trees. They say, 'Don't touch this tree because I know the old guy who planted this tree.' They have a respect for the old guy, and they translate this respect to the tree, the forest. It's quite a feeling. These guys are very careful. When they log, they know exactly what they're doing.

"The forester has not only to reason out which tree to cut; he has to mark which direction it has to fall. He has to do it in such a way that his first concern is the replacement, not to lose the growing stock. What can he substitute in its place? He releases some younger trees, which then will grow faster. This is the process—but sometimes he makes mistakes. Then the faller will immediately spot the mistake. He'll recognize which way the tree will go, and, if that differs from what the forester had in mind, the faller can halt the process right away. He can say, 'I don't want to cut the tree this way.' Then there is an investigation made to determine which way it will go. There's a lot more interaction between the boss and the workers.

"The forester is supposed to be able to do it himself if he has to. You can't ask the faller to do it that way if you can't do it yourself. In Europe, tree falling is part of the curriculum for professional foresters. They study the theory, the technique, the safety rules, the forces involved. In Europe, you compute the force. You're supposed to be able to *compute* the force for a tree of this volume when it hits the ground. That way you get an idea of how much breakage you will get. I tried to introduce tree falling into the curriculum here, but I got into all kinds of trouble because of liability. It's too risky. But in Europe you have two years on the apprentice level, and you have the old logger who's going to teach you how to do it. By that time you already have the knowledge of what is technically feasible. You have both book knowledge and field knowledge. And the forester is supposed to know a lot of other things. When the tractor stops, you have to be able to pull the plug and figure out what is wrong. It's all part of the apprentice period.

"In the Couvet forest, the forester and the logger both take pride in how carefully they can bring down a tree. It's built right into the system. You have to save your regeneration or else you lose your prestige, your respect. It's your professional standard. Certain trees that are too big-crowned are earmarked, and then the faller will have to climb the tree. He gets extra money for that. He will saw off the branches and actually lower the branches by rope, so they don't do any damage. He doesn't let them fall lest they break some regeneration. Then once the branches are off, he will drop the tree like a pencil into the stand.

"In Europe the hierarchy in the forest service is quite different from what it is in this country. The goal of each forester is to manage his district. That becomes then his home, his source of pride. He uses all his expertise to develop it, to plan the roads, and so on. He is not transferred. He will stay there. He is only transferred when he gets an administrative position—and he usually resists that.

"Foresters are actually kings of their forests. They usually have the right to be buried in their forests. They have the right to put up trespass signs. They are in control of the total forest: the wildlife, the water, the trees. They have the absolute hunting rights. You don't have the right to bear arms in Europe, so you have to have a hunting license. You hunt by invitation. You are invited by the forester to join the hunting party. They hunt on certain days, and they hunt only for certain things. They say, 'Okay, we want to kill three deer: the one with the nick in the ear, the one with this thing, and the one with that.' They try to keep the deer population in balance. They try to figure out how many deer are in the forest. If there is too much pressure on one area, they invite people to hunt those specific deer.

"In Switzerland, the foresters are still elected. Traditionally, if a guy is doing his job he is automatically elected. But he still appears on the ballot. It's a position of public trust. It gives you a status. It gives you power and prestige. And you're supposed to act on the public's behalf.

"The forester in Europe is also the manager of the business. He sets policy and determines what is happening. In this country, the forester is not in charge of the business. That's left to the corporate economist. The forester is like a technician without a professional goal. So the problem is that in many cases there is no leadership. There's no continuity.

"Under all-age management, you have to maintain continuity. When you skid your logs out, you use a constant road. A tree planter has to know that this is the direction the logs get skidded out; he can't go planting his trees in the skid trail. So he has to know his compartment.

"The road network has to be well planned. The roads don't just follow straight lines; they're curved to fit the landscape with the least amount of cut. You have to be gentle with the roads. You don't need a freeway; but you need access to the forest, because the whole area is intensively developed and managed. The roads are

Preparing to fall timber: learning garden forestry in Switzerland

small, but they serve a lot of purposes: they are used for fire protection; they provide browse for deer, for wildlife; they let light into the stand for regeneration; they're hiking trails in the summer and cross-country ski trails in the winter; and they provide access to the stand for maintenance or measurement.

"But the equipment is confined to the roads. The main tractor is not allowed to go into the stand *per se*. It can only travel the skid trail. The cable is strung out through the trees to the log, and you winch it in. You stack the logs along the skid trail in piles. The people that walk along the skid trail can see the logs and turn in their bids. You buy maybe five, six piles, or maybe only one. The piles are sufficiently small so that everybody can bid on them.

"The tractor travels the skid road backward, so it doesn't have to turn around and create a tremendous circle. The road is too narrow for that. Also, in Switzerland they do it over snow, so there's no damage to the soil. They wait until the snow hardens a foot or more, and then they skid over the snow on sleds.

"The equipment is generally smaller in Europe than it is in this country. Most of the equipment traditionally developed from the farm. They use a lot of horses in the woods.

"Many foresters have come from the farms. They go off to the university, and then they come back to their homes. They work right along with their fathers. A tradition develops of working on the same land for generation after generation. There is more interest in keeping the land productive.

———— ,, ————

The Swiss system is, in a word, *personalized*. The foresters and forest workers are dealing with individual trees on limited forest land, not with large, anonymous tracts of raw timber. There is no financial reward for liquidating resources; indeed, the professional standards are based on how well regeneration can be accomplished. The forest is seen as a complete entity that grows timber, nourishes wildlife, stabilizes hillsides, provides water, and serves the recreational needs of human beings. The forester is the caretaker—but not the owner—of this entity. He is a "ranger" in the old-fashioned sense: a keeper of the woods, a steward of the land. As a public servant and an elected official, he is charged with the task of maintaining a healthy, balanced, and productive forest.

The Swiss system as practiced in Couvet forest is a living example of holistic forestry. The system is not based upon voluminous restrictions imposed by bureaucratic structures; it did not come about by the incessant harangues of environmentalists. Instead, it evolved from a combination of local tradition and a deeply felt need. The need was to maximize sustained-yield production on a small amount of land, to manage the land to its optimum potential, to waste nothing. The tradition was the farmers' bonding to the land over successive generations, and their respect for the inherent value of trees. Historically, the Swiss have treated the forest as part of the public domain, the collective heritage of the local inhabitants.

People in the Forests: Can We Do It Here?

Could the Swiss system be translated into an American form? Tradition, of course, cannot be exported. The forest in this country is not always seen as a collective heritage, nor are the trees given the individualized attention they receive in Switzerland. But we certainly feel the need: the forest must be sustained, and people must find work.

At home and work: the Swiss forester's stewardship

If our needs are to be fulfilled, however, we will somehow have to pay the price. The costs of liquidating our forests have not been met. We have tried to cut corners by simplifying the ecosystem and shortening the growing cycle, but we now suspect that this cannot be done without many sacrifices: lost Iobs, diminished wildlife habitat, inferior wood.

On the surface, the direction we must take seems clear: employ the people to tend the woods more carefully—and be willing to pick up the tab. But even if we can find mechanisms to share the costs appropriately, the notion that human needs can be fulfilled while simultaneously maintaining healthy forests requires close and continuous scrutiny. How much wood can we remove from an ecosystem before we significantly alter the nutrient cycling? As we tinker with natural systems, we must retain a certain element of humility: we don't always know what we are doing. In the words of one concerned observer: "We did not design the forest, so we do not have a blueprint, parts catalog, or maintenance manual with which to understand and repair it."[2]

The lush forests of the Pacific Northwest have been around for thousands of years. With the retreat of the glaciers at the end of the last ice age, climatic and soil conditions gradually became suited for the creation of "the classic coniferous forests of the world."[3] Euro-American civilization, by contrast, has been a forceful presence in this region for barely 150 years, far less than the age of many of the trees. Our attempts to practice any type of forestry in the Northwest date back only about

50 years, less than the time it takes a single generation of Douglas-firs to reach maturity. We have placed the forests under tremendous stress during our short reign, and even now we don't understand all the consequences of our actions.

We would be well advised, therefore, to proceed slowly and cautiously as we try to develop a more sustainable human presence in the woods. We need to operate more within the time frame of the forest itself. When we try to speed up the meandering pace of natural processes, we place the forest at risk—and we simultaneously threaten to destroy the resource base which supports our own economic activity.

In the heat of the moment, concerned as we are with our pressing and immediate needs, how can we keep a safe and sane perspective? How can we get back to the big picture? How can we see the forest for the trees?

At the very least, we must learn how the forest really works before trying to manipulate it to our own advantage. We must understand the consequences of our actions—not only of our harvesting activities, but of our attempts to restore the health of ecosystems. For all our good will, how do we know we won't be doing more harm than good? We once thought, for instance, that we were doing right by the forests when we put out all the little fires, but suddenly we realize that our well-intentioned efforts have only served to increase the likelihood of major catastrophes. In the 1970s and early '80s, we tried to remove all the fallen logs from the streams in order to facilitate the movement of fish; today, we are putting some of the logs back in the water to provide riffles, deep pools, and habitat for invertebrates which serve as food for the fish. Even in Switzerland, where the people tended their forests so carefully for generations, the efforts to keep the woods tidy have backfired: the steady removal of woody debris has had a significant impact on the nutrient capital of the soil.

Our obvious need for knowledge can be filled only by more and better research. But the problem with forest research, as with most of our management activities, is an incongruity of time scale. We want the answers *now*, while forest ecosystems simply do not behave in such a hasty manner. In order to understand what is really happening in the woods, we will have to slow down. Obviously, we cannot afford to retreat altogether from being participants within the forests—or from using forest products. But our attitude should reflect our growing awareness that we are a part of the ecosystem, not above it, and that the pace of the forest is slower than our own. Although we want it all now, these things do take time.

Mark Harmon

Assistant Professor, Oregon State University

"When the people at H. J. Andrews came up to me and said they needed somebody to design a 200-year study on log decomposition, it looked like the opportunity of a lifetime. I've always been interested in decomposition. I know it sounds strange, but as a kid I used to play in our compost pile, look at the stuff disappear. I guess it just rubbed off on me. Mainly we were digging in the compost to find worms for fishing, but I do remember wondering: How does this stuff turn into soil?

"We're trying to look at the factors that control the decomposition of dead trees, and the way that they interact with the forest ecosystem. This is a real long-term process. We actually started by going out and trying to find trees that died at different ages; that's where we learned a lot about the function of dead trees. But there are lots of things you can't control in that kind of study, so we ended up designing a study where we put out things right from the beginning. We knew what they looked like at the start, and then we could watch them develop through time. We could control not only the species but the size of the trees. Some of them might be rotten to begin with, while some of them might be sound, and now we can control for that.

"We put out logs that are about twenty feet long and about two feet in diameter. We selected the trees ahead of time so they'd be the right size and species. Then we had people cut them down and bring them into the experimental areas in trucks. We made sure that the sites were all pretty similar, and we made little places for them to rest so that they'd all be on the ground. Then every once in awhile we go out and cut some of the pieces up and thoroughly examine them: where the decay has spread to, what kind of insects live in them, what kind of fungi are growing on them—all kinds of little processes. We keep examining these things periodically to see how they change.

"For the first ten years, we take most of our measurements every year. We also do a few measurements monthly, because we're interested in the seasonal changes. After that we'll go in every couple of years, and then after twenty years we'll just be looking at it about every ten years. Whenever someone gets interested and wants to look at it, it'll be there. We've made maps so people will be able to find them, and we store all the data in the computer. We have it all organized so there'll be a good record for people to go back to.

"They've been out eight years so far, so they're still in the early stages. We've got 192 years left to see what happens. I know it sounds bizarre to put out something for 200 years. Well, people have done this for just a few years, but then they end up saying: 'I wonder what would happen if it was longer.' So we decided: 'Let's do it right. Let's plan ahead, and leave out enough for other people to look at in the future.' Two hundred years was chosen just as a rough estimate of how long it might take for the logs to decompose. Probably some pieces will still be there in 200 years, some little parts of them.

"We're interested not only in the long-term process of decay but in nutrient release. That's not very well understood for wood. There're lots of ideas, but our data is showing that those ideas do not always explain what's really happening. There are mechanisms for release that people haven't thought about. When people have analyzed nutrient release in wood before, they usually used sawdust. But sawdust is different from a log. A log is a separate entity; you have these things going on inside a log, and then the mushrooms that are produced fall off the log and get added to the soil. These are things that don't happen if you're just looking at sawdust.

"I think we'll be constantly surprised until we start looking at this stuff more carefully. This is the first time these processes have really been watched systematically. As Yogi Berra said (or maybe it was Casey Stengel), 'Sometimes, if you look at something, you might see it.' That's what we're doing. People have spent a lot of time speculating about what would happen; now we're looking at it and realizing that those theories don't really work for this system.

"In many cases, ecosystems just don't make much sense until you look at them over a long period. They're too complex. Somebody called it 'the invisible present.' What they meant by that is sometimes you don't know what something means until you expand your time horizon. We know that the average temperature in Corvallis this year was 15 degrees centigrade. But what the hell does that mean? Well, it could have been a very cold year, or it could have been part of a global warming trend. You can't tell what that 15 degrees means until you start expanding your time horizon. That's why the present is invisible; it really doesn't take on any meaning until you compare it to different points in time. And as you expand out further and further, it takes on different meanings. When you take just the short-term view, you're always dealing in the invisible present. You don't get a real sense of what's going on.

"Short-term studies in forestry often give very misleading results; you can't always tell what the long-term effects will be. A classic example is the effect of a shrub like snowbush or *Ceanothus*, which is a nitrogen-fixer. Early on, it's competing with the crop trees, shading them and seemingly hurting the production. So you go out there and cut them all down to get your seedlings to grow faster. But this shrub does a couple of things. It reduces the stocking level so that when the trees do come in, they're not overly dense. It also adds nitrogen to the soil. In the short term, if you remove them, the production can go up. But in the long term, the crop trees will eventually kill off the brush, and when they do, the soil will be better quality and the trees won't be competing with each other. In a hundred years, it's more productive if you don't wipe out the *Ceanothus*.

"Our project is part of what they call LTER—Long-Term Ecological Research. It's a special program developed within the last ten years that acknowledges that for many of these ecosystem processes, you don't really learn much by looking in the short term. You have to look at things in their own time scale. We're the only ones looking at logs, but there are 18 different sites, with hundreds of people doing long-term studies on populations, disturbances, decay, and so on. Here at H. J. Andrews, there are 20 or more Ph.D. scientists working on it; that's true of the other sites too. They have a couple of sites up in Alaska, in Massachusetts, Puerto Rico, Antarctica. They try to cover different ecological zones.

"My personal plan is to try and stay with this project my whole professional career, and then pass it on. I'm forty years old, so I probably have a few more decades left in me. I'm not planning to make it for the whole 200 years, but that's all right. You have to pretend that the future is going to happen. When we started this project, things weren't looking too good between the Soviet Union and the U.S. Some people were saying, well, the world's going to end anyway, so why bother? You look at it now and all that's changed. You never can tell, so you just have to take the risk that somebody will still be around.

"We've actually inherited some data sets from the twenties, and even earlier. The people who set these things up are long gone, but we still use their information, take measurements on their studies, keep track of what's going on. What they set up has been invaluable. I've seen the other end of it, so I have the feeling that there will be somebody paying attention. I certainly hope so.

——— ” ———

Notes

Chapter 1

The interviews with Sid Stiers and Ernest Rohl were conducted in 1980.

1. "Reminiscences of Mendocino," *Hutchings California Magazine,* October 1858.

2. Ron Finne, *Natural Timber Country* (16-mm film).

3. Jess Larison, 1979: personal communication.

4. Gifford Pinchot, *The Fight for Conservation* (Seattle: University of Washington Press, 1967), pp. 14–15.

5. 16 *United States Code,* p. 475.

6. Shirley W. Allen and Grant W. Sharpe, *Introduction to American Forestry* (New York: McGraw-Hill, 1960), p. 331.

7. Finne, *Natural Timber Country.*

8. Jess Larison, 1979: personal communication.

9. USDA Forest Service, *Douglas-fir Supply Study* (Portland: Pacific Northwest Forest and Range Experiment Station, 1969).

10. Finne, *Natural Timber Country.*

11. Ibid.

Chapter 2

The interviews with Bill Hagenstein, Philip F. Hahn, and Robert Barnum were conducted in 1980.

1. Herman H. Chapman and Walter H. Meyer, *Forest Valuation* (New York: McGraw-Hill, 1947), p. 88.

2. *American Forests,* 84 (10): October, 1978, p. 15.

3. Nancy Wood, *Clearcut: The Deforestation of America* (San Francisco: The Sierra Club, 1971), p. 17.

4. Gifford Pinchot, *Breaking New Ground* (New York: Harcourt Brace, 1947). Cited in Daniel R. Barney, *The Last Stand: Ralph Nader's Study Group Report on the National Forests* (New York: Grossman, 1974), pp. 64–65.

5. Wood, *Clearcut,* p. 30.

6. Ibid., p. 62.

7. Ibid., p. 71.

8. Philip F. Hahn, "The Effect of Animal Damage on Volume Growth," (Eugene: Georgia-Pacific Corp., 1978).

9. *Report from the California Redwood Association* (Eureka: Redwood Region Logging Conference, 1978).

10. P. Bettinger, J. Sessions, and L. Kellogg, "Potential Timber Availability for Mechanized Harvesting in Oregon," *Western Journal of Applied Forestry* 8 (1993): 11–15.

11. *Forest Industries,* November 1978, p. 60.

Chapter 3

The interviews with Gordon Robinson and Fred Behm were conducted in 1980; the first half of the interview with Nat Bingham was conducted in 1980, the second half in 1993.

1. Ron Finne, *Natural Timber Country* (16-mm film).

2. Esteban de la Puente 1980: personal communication.

3. R. W. Stark, "The Entomological Consequences of Even-Age Management," *Even-Age Management.* Richard K. Hermann and Denis P. Lavender, eds. (Corvallis: Oregon State University Press, 1973).

4. John R. Parmeter, Jr., "Ecological Considerations in Even-Age Management: Microbiology and Pathology," *Even-Age Management.* R. K. Hermann and D. P. Lavender, eds. pp. 122–123.

5. Samuel A. Graham, *Forest Entomology* (New York: Reinhold, 1952), p. 66.

6. T. D. Schowalter, "Ecological Strategies of Forest Insects," *New Forests* 1 (1986): 57–66.

7. Frederick E. Smith, "Ecological Demand and Environmental Response," *Journal of Forestry,* December 1970, p. 755.

8. R. F. Huttl, "Forest Decline and Nutritional Disturbances," in *Forest Site Evaluation and Long-Term Productivity.* D. W. Cole and S. P. Gessel, eds. (Seattle: University of Washington Press, 1988), pp. 180–186.

9. M. P. Amaranthus, J. M. Trappe, and R. J. Molina, "Long-Term Productivity and the Living Soil," *Maintaining the Long-Term Productivity of Pacific Northwest Ecosystems.* D. A. Perry, R. Meurisse, B. Thomas, R. Miller, J. Boyle, J. Means, C. R. Perry, and R. F. Flowers, eds. (Portland: Timber Press, 1989), p. 40.

10. Ibid., p. 42.

11. J. M. Trappe, "Selection of Fungi for Ectomycorrhizal Inoculation in Nurseries," *Annual Review of Phytopathology* 15 (1977): 203–222.

12. C. Maser, J. M. Trappe, and R. Nussbaum, "Fungal–Small Mammal Interrelationships, With Emphasis on Oregon Coniferous Forests," *Ecology* 59 (1978): 799–809.

13. Roy R. Silen, "The Care and Handling of the Forest Gene Pool," *Pacific Search,* June 1976, p. 9.

14. F. J. Swanson, J. L. Clayton, W. F. Megahan, and G. Bush, "Erosional Processes and Long-Term Site Productivity," in *Maintaining the Long-Term Productivity of Pacific Northwest Ecosystems.* D. A. Perry *et al.,* eds., p. 77.

15. S. N. Little and J. L. Ohmann, "Estimating Nitrogen Loss from Forest Floor during Prescribed Burns in Douglas-fir/Western Hemlock Clearcuts," *Forest Science* 34 (1988): 152–164.

16. D. H. McNabb and K. Cromack, Jr., "Effects of Prescribed Fire on Nutrients and Soil Productivity," in *Natural and Prescribed Fire in Pacific Northwest*

Forests. J. D. Walstad, S. R. Radosevich, and D. V. Sandberg, eds. (Corvallis: Oregon State University Press, 1990), pp. 125–142.

17. M. C. Feller, J. P. Kimmins, and K. M. Tsze, "Nutrient Losses to the Atmosphere during Slashburns in Southwestern British Columbia," *Proceedings, 7th Conference on Fire and Forest Meteorology* (Boston: American Meteorological Society, 1983), pp. 123–135.

18. R. E. Martin, "Prescribed Burning for Site Preparation in the Inland Northwest," *Tree Planting in the Inland Northwest* (Wash. State Univ. Conf., 1976).

19. A. E. Harvey, M. J. Larsen, and M. F. Jurgensen, "Comparative Distribution of Ectomycorrhizae in Soils of Three Western Montana Forest Habitat Types," *Forest Science* 25(1979): 350–360.

20. C. T. Dyrness, *Effect of Wildfire on Soil Wettability in the High Cascades of Oregon.* Research Paper PNW-202. (Portland: Pacific Northwest Forest and Range Experiment Station, USDA Forest Service, 1976).

21. C. T. Youngberg, A. G. Wollum, and W. Scott, "*Ceanothus* in Douglas-fir Clearcuts: Nitrogen Accretion and Impact on Regeneration," in *Symbiotic Nitrogen Fixation in the Management of Temperate Forests.* J. C. Gordon, C. T. Wheeler, and D. A. Perry, eds. (Corvallis: Oregon State University, 1979), pp. 224–233.

22. M. Newton, B. A. El Hassan, and J. Zavitkovski, "Role of Red Alder in Western Oregon Forest Succession," in *Biology of Alder.* J. M. Trappe, J. F. Franklin, R. F. Tarrant, and G. M. Hansen, eds. (Portland: Pacific Northwest Forest and Range Experiment Station, USDA Forest Service, 1968), pp. 73–84; Robert F. Tarrant, K. C. Lu, W. B. Bollen, and J. F. Franklin, *Nitrogen Enrichment of Two Forest Ecosystems by Red Alder,* Research Paper PNW-76 (Portland: Pacific Northwest Forest and Range Experiment Station, USDA Forest Service, 1969); K. Cromack, C. Delwiche, D. H. McNabb, "Prospects and Problems of Nitrogen Management Using Symbiotic Nitrogen Fixers," in *Symbiotic Nitrogen Fixation.* J. C. Gorgon *et al.,* eds., pp. 210–223.

23. Robert F. Tarrant, K. C. Lu, W. B. Bollen, and C. S. Chen, *Nutrient Cycling by Throughfall and Stemflow Precipitation in Three Coastal Oregon Forest Types,* Research Paper PNW-54 (Portland: Pacific Northwest Forest and Range Experiment Station, USDA Forest Service, 1968); R. F. Tarrant *et al., Nitrogen Enrichment by Red Alder.*

24. R. F. Tarrant *et al., Nitrogen Enrichment by Red Alder;* Newton et al., "Role of Red Alder."

25. M. P. Amaranthus *et al.,* "Long-Term Productivity and the Living Soil," *Maintaining the Long-Term Productivity of Pacific Northwest Ecosystems.* D. A. Perry *et al.,* eds., p. 45.

26. S. L. Borchers and D. A. Perry, "Growth and Ectomycorrhiza Formation of Douglas-fir Seedlings Grown in Soils Collected at Different Distances from Pioneering Hardwoods in Southwest Oregon Clear-cuts," *Canadian Journal of Forest Research* 20 (1990): 712–721.

27. A. E. Harvey, M. J. Larson, M. F. Jurgensen, and J. A. Schlieter, *Distribution of Active Ectomycorrhizal Short Roots in Forest Soils of the Inland Northwest:*

Effects of Site and Disturbance, General Technical Report INT-374 (Ogden: Intermountain Research Station, USDA Forest Service, 1986).

28. M. P. Amaranthus *et al.* "Long-Term Productivity and the Living Soil," *Maintaining the Long-Term Productivity of Pacific Northwest Ecosystems.* D. A. Perry *et al.,* eds., p. 48.

29. When whole trees, rather than just the trunks, are harvested, twice as much nitrogen leaves the forest. McNabb and Cromack, "Effects of Prescribed Fire," in *Fire in Pacific Northwest Forests.* J. D. Walstad *et. al.,* eds.

30. Joseph Kittredge, *Forest Influences* (New York: McGraw-Hill, 1948).

31. D. D. Harris, "Hydrologic Changes after Clearcut Logging in a Small Coastal Oregon Watershed," *J. Res. U.S. Geol. Survey* 1: 487–491. Cited in Marvin Dodge, L. T. Burcham, Susan Goldhaber, Bryan McCulley, and Charles Springer, *An Investigation of Soil Characteristics and Erosion Rates on California Forest Lands* (Sacramento, Calif., 1976), p. 25.

32. Gerald Myers, 1979: personal communication.

33. W. D. Ellison, "Soil Erosion," *Soil Sci. Soc. Amer. Proc.* 12 (1947): 479–484. Cited in M. Dodge *et al., Erosion Rates,* p. 13.

34. Robert N. Coats, "The Road to Erosion: A Cautionary Tale," *Environment,* January–February 1978.

35. R. L. Fredriksen, *Erosion and Sedimentation Following Road Construction and Timber Harvest on Unstable Soils in Three Small Western Oregon Watersheds.* Research Paper PNW-104. (Portland: Pacific Northwest Forest and Range Experiment Station, USDA Forest Service, 1970).

36. C. T. Dyrness, *Mass Soil Movements in the H. J. Andrews Experimental Forest.* Research Paper PNW-42 (Portland: Pacific Northwest Forest and Range Experiment Station, USDA Forest Service, 1967).

37. M.P. Amaranthus, R. M. Rice, N. R. Barr, and R. R. Ziemer, "Logging and Forest Roads Related to Increased Debris Slides in Southwest Oregon," *Journal of Forestry* 83 (1985): 229–233.

38. Jack S. Rothacher and Thomas B. Glazebrook, *Flood Damage in the National Forests of Region 6* (Portland, 1968). Cited in M. Dodge *et al., Erosion Rates,* p. 39.

39. E. C. Steinbrenner and S. P. Gessel, "The Effect of Tractor Logging on Physical Properties of Some Forest Soils in Southwestern Washington," *Soil Sci. Soc. Amer. Proc.* 19 (1955): 372–376.

40. Ibid.

41. R. L. Fredriksen and R. D. Farr, "Soil, Vegetation, and Watershed Management," *Forest Soils of the Douglas-fir Region.* P. E. Heilman, H. W. Anderson, and D. M. Baumgartner, eds. (Pullman: Washington State University Extension Service, 1979), 231–260.

42. J. D. Hall and R. L. Lantz, "Effects of Logging on the Habitat of Coho Salmon and Cutthroat Trout in Coastal Streams," *Symposium on Salmon and Trout in Streams.* T. G. Northcote, ed. (Vancouver, B.C., 1969).

43. R. L. Lantz, *Guidelines for Stream Protection in Logging Operations* (Portland, Oreg.: Oregon State Game Commission, 1971).

44. *An Environmental Tragedy* (Sacramento, Calif.: California Dept. of Fish and Game, 1971).

45. R.N. Coats, "Road to Erosion."

46. N. Wood, *Clearcut,* p. 70.

47. Charles F. Wilkinson, *Crossing the Next Meridian* (Washington, D. C.: Island Press, 1992), p. 160.

48. J. F. Franklin, D. A. Perry, T. D. Schowalter, M. E. Harmon, A. McKee, and T. A. Spies, "Importance of Ecological Diversity in Maintaining Long-Term Site Productivity," in *Maintaining the Long-Term Productivity of Pacific Northwest Ecosystems.* D. A. Perry *et al.,* eds., p. 93.

Chapter 4

The interviews with Gerald Myers and Rick Koven were conducted in 1980; the interview with Peggy Iris was conducted in 1993.

1. See interview with Gordon Tosten, Chapter 9.

2. For a discussion of the nitrogen-fixing properties of red alder, see Chapter 3.

3. R.F. Tarrant *et al., Nitrogen Enrichment by Red Alder;* M. Newton *et al.,* "Role of Red Alder."

4. H. Miller, "Long-Term Effects of Application of Nitrogen Fertilizers on Forest Sites," in *Forest Site Evaluation and Long-Term Productivity.* D. W. Cole and S. P. Gessel, eds., pp. 97–106.

5. R. E. Miller and M. D. Murray, "The Effects of Red Alder on Growth of Douglas-fir," in *Utilization and Management of Alder,* General Technical Report PNW-70 (Portland: Pacific Northwest Forest and Range Experiment Station, USDA Forest Service, 1978), pp. 283–304; W. A. Atkinson and W. I. Hamilton, "The Value of Red Alder as a Source of Nitrogen in Douglas-fir/Alder Mixed Stands," in *Utilization of Alder,* pp. 337–351.

6. The economic viability of various schemes for mixing alder with Douglas-fir is explored in R. E. Miller and M. D. Murray, "Fertilizer versus Red Alder for Adding Nitrogen to Douglas-fir Forests of the Pacific Northwest," in *Symbiotic Nitrogen Fixation.* J. C. Gordon *et al.,* eds. pp. 356–373; and W. A. Atkinson and W. I. Hamilton, "Value of Red Alder," in *Utilization of Alder,* pp. 337–351.

7. For a summary of these reports, with bibliographical references, see John W. Warnock and Jay Lewis, *The Other Face of 2,4-D* (Penticton, B.C.: South Okanagan Environmental Coalition, 1978), pp. II-1 to II-11.

8. *San Francisco Chronicle,* 9 April 1977.

9. Wilbur P. McNulty, Jr., "Testimony on the Effects of Tetra-Dioxin on Primates," Citizens Against Toxic Sprays v. U.S. Dept. of Agriculture, U.S. District Court, District of Oregon, 1976.

10. Jack Anderson, "Defoliating America," *San Francisco Chronicle,* 24 April 1978.

11. *10-Year Forest Development Plan and Supplemental Forest Development Plan for 1978–1988 Cutblocks* (Hoopa, CA: Hoopa Valley Natural Resource Department, Forestry Division, 1990).

12. J. D. Walstad, M. Newton, D. H. Gjerstad, "Overview of Vegetation Management Alternatives," *Forest Vegetation Management for Conifer Production.* J. D. Walstad and P. J. Kuch, eds. (New York: John Wiley, 1987).

13. C. Click, J. N. Fiske, J. Sherlock, and R. Wescom, "Alternatives to Herbicides," *Proceedings, Tenth Annual Forest Vegetation Management Conference* (Redding: Forest Vegetation Management Conference, 1988); G. O. Fiddler and P. M. McDonald, "Manual Release Contracting: Production Rates, Costs, and Future," *Western Journal of Applied Forestry* 5 (1990): 83–85.

14. Groundwork, Inc., *Willamette Brush Control Study Project* (Eugene, 1978).

15. J. Franklin *et al.*, "Importance of Ecological Diversity."

16. T. F. Hughes, J. C. Tappeiner, and M.Newton, "Relationship of Pacific Madrone Sprout Growth to Productivity of Douglas-fir Seedlings and Understory Vegetation," *Western Journal of Applied Forestry* 5 (1990): 20–24.

17. A. P. Amaranthus and D. A. Perry, "Interaction Effects of Vegetation Type and Pacific Madrone Soil Inocula on Survival, Growth, and Mycorrhiza Formation of Douglas-fir," *Canadian Journal of Forest Research* 19 (1989): 550–556. The positive effects of the madrone soil worked only on sites occupied primarily by manzanita; on sites occupied by grasses or white oak, it had little or no effect. This experiment demonstrated an interesting ecological interrelationship between madrone, manzanita, and Douglas-fir.

Chapter 5

The interview with Bud McCrary was conducted in 1980; the interview with Orville Camp was conducted in 1993.

1. D. M. Moehring and I. K. Rawls, "Detrimental Effects of Wet Weather Logging," *Journal of Forestry,* 68 (1970): 166–167.

2. P. M. McDonald, W. A. Atkinson, and D. O. Hall, "Logging Costs and Cutting Methods in Young Growth Ponderosa Pine in California," *Journal of Forestry,* 67 (1969): 109–113.

3. USDA Forest Service, *Final Environmental Statement and Renewable Resource Program, 1977–2020,* Report to Congress, 1976.

4. Ted Blackman, "Computer Tells Foresters How Specific Practices Would Work," *Forest Industries,* January 1979.

5. Until the mid-1970s, California had an *ad valorum* timber tax; the value of stands over the age of 40 was included in the calculations of property taxes.

6. Promotional brochure: "Controlled Falling . . . A Lean Toward Higher Profits," Silvey Precision Chain Grinder Co., Eagle Point, OR.

7. Ibid; see also Dave Burwell, unpublished paper on controlled falling, Springfield, OR.

8. "The Uncommon Forest Management Plan," *Timber/ West,* February 1978. With three other foresters, Smith managed about 40,000 acres until his death in 1989.

9. Ibid., p. 26.

10. Ibid.

11. The economic concept alluded to here will be discussed in Chapter 7.

12. Don Minore, "Shade Benefits Douglas-fir in Southwestern Oregon Cutover Area," *Tree Planters' Notes* 22 (1): Feb., 1971, pp. 22–23.

13. S. W. Childs and L. E. Flint, "Effect of Shadecards, Shelterwoods, and Clearcuts on Temperature and Moisture Environments," *Forest Ecology and Management,* 18 (1987): 205–217.

14. See J. Franklin *et al.,* "Importance of Ecological Diversity."

Chapter 6

The interviews with Joe Miller and Joan and Joaquin Bamford were conducted in 1980; the interview with Wally Budden and Terry and Vicki Neuenschwander was conducted in 1993.

1. S. W. Childs, S. P. Shade, D. W. R. Miles, E. Shepard, and H. A. Froehlich, "Soil Physical Properties: Importance to Long-Term Forest Productivity," in *Maintaining the Long-Term Productivity of Pacific Northwest Ecosystems.* D. A. Perry *et al.,* eds., p. 62.

2. R. N. Coats, "Road to Erosion."

3. J. A. Helms and C. Hipkin, "Effects of Soil Compaction on Tree Volume in a California Ponderosa Pine Plantation," *Western Journal of Applied Forestry* 1(1986): 121–124. These figures are confirmed by other related studies. The leader growth on Douglas-fir seedlings in western Oregon showed a 43% reduction on tractor trails; the growth on 8-year-old Douglas-firs showed a 57% reduction. [C. T. Youngberg, "The Influence of Soil Conditions Following Tractor Logging on the Growth of Douglas-fir Seedlings," *Soil Sci. Soc. Amer. Proc.* 23 (1959): 76–78.] According to another study, 26-year-old loblolly pines that grew back on old roadways had only one-half the volume of wood of their immediate neighbors. [Henry A. Froehlich, "The Impact of Even-Age Forest Management on Physical Properties of Soil," *Even-Age Management.* R. K. Hermann and D. P. Lavender, eds., pp. 199–220.]

4. S. Wert and B. R. Thomas, "Effects of Skid Roads on Diameter, Height, and Volume Growth in Douglas-fir," *Soil Sci. Soc. of Amer. Journal* 45 (1981): 629–632.

5. C. T. Dyrness, "Soil Surface Disturbance Following Tractor and High-Lead Logging in the Oregon Cascades," *Journal of Forestry* 63 (1965): 272–275.

6. G. O. Klock, "Impact of Five Postfire Salvage Logging Systems on Soils and Vegetation," *J. Soil and Water Conservation* 30 (2): March–April 1975, pp. 78–81.

7. M. Dodge *et al., Erosion Rates,* p. 60.

8. Interview with Faye Stewart, in Ray Raphael, *Tree Talk: The People and Politics of Timber* (Covelo, CA: Island Press, 1981), pp. 144–150.

9. Steve Conway, *Logging Practices: Principles of Timber Harvesting Systems* (San Francisco: Miller Freeman Publications, 1982). For earlier figures, see also Doyle Burke, "Helicopter Logging: Advantages and Disadvantages Must Be Weighed," *Journal of Forestry* 71 (1973): 574–576.

10. S. Conway, *Logging Practices.*

Chapter 7

The interviews with Bill Bertain and Jim Rinehart were conducted in 1993.

1. Richard W. Haynes, *An Analysis of the Timber Situation in the United States: 1989–2040,* General Technical Report RM 199 (Fort Collins: Rocky Mountain Forest and Range Experiment Station, USDA Forest Service, 1989).

2. Ibid., p. 55.

3. J. H. Beuter and K. N. Johnson, "Economic Perspectives on Maintaining the Long-Term Productivity of Forest Ecosystems," *Maintaining the Long-Term Productivity of Pacific Northwest Ecosystems.* D. A. Perry *et al.,* eds., p. 222.

4. Site class II, site index 170.

5. Steven Calish, Roger D. Fight, and Dennis E. Teeguarden, "How do Nontimber Values Affect Douglas-fir Rotations?" *Journal of Forestry* 76 (1978): 217–221. "McArdle and Meyer's (1930) normal yield table for Douglas-fir, site index 170, was used as the basis for calculating a timber-only rotation. Yields, expressed in thousand cubic feet per acre, were reduced by a factor of 0.94 to account for deer browsing on early growth. For lack of better data, we assumed that rotations may be repeated without affecting site productivity. Stumpage revenue and initial costs were expressed in real dollars. Stumpage was valued from $700 per thousand cubic feet for 10-inch trees to about $875 for 21-inch trees. Regeneration cost was set at $75 per acre, and it was assumed that successful regeneration occurs immediately following harvest. For the data enumerated, and a 5 percent rate of interest, the best economic rotation for timber production is 36 years, resulting in an SEV (soil expectation value) of $643 per acre. In contrast, if culmination of mean annual increment is used, the rotation is 64 years and SEV is $329 per acre" (p. 218).

6. Site class II, site index 170.

7. Yield tables taken from T. F. Arvola, *California Forestry Handbook* (Sacramento, 1978), pp. 215–217.

8. Bill Wilkinson, Natural Resources Department, Forestry Division, Hoopa Valley Tribal Council: personal communication, 1993.

9. W. A. Atkinson and W. I. Hamilton, "Value of Red Alder."

10. James A. Rinehart and Paul S. Saint-Pierre, *Timberland: An Industry, Investment, and Business Overview* (Boston: Hancock Timber Resource Group, 1991), p. 10.

11. Michael Shnayerson, "Redwood Raider," *Houston Metropolitan,* January 1992, p. 44.

12. Forest and Rangeland Resources Assessment Program (FRRAP), *California's Forests and Rangelands: Growing Conflict over Changing Uses* (Sacramento: California Department of Forestry and Fire Protection, 1988), p. 163.

13. Seth Zuckerman, "Old Forestry: A Visit to California's Collins Almanor Forest," *Sierra,* March/April 1992, p. 44.

14. FRRAP, *California's Forests and Rangelands,* p. 154.

Chapter 8

The interview with Raymond Cesaletti was conducted in 1980; the interviews with Rich Fairbanks and John Berry were conducted in 1993.

1. R. W. Haynes, *Timber Situation in the United States.*

2. C. F. Wilkinson, *Crossing the Next Meridian,* p. 116.

3. See above, Chapter 1.

4. 16 *United States Code,* pp. 528–531.

5. Robert E. Wolf, "Ignoring the Letter of the Law," *Inner Voice* 1(2): Fall 1989. Ironically, the NFMA originated as an industry response to a court ruling that appeared to ban clearcutting in the National Forests. The 1897 Organic Act, in establishing the guidelines for the Forest Service, had stated specifically that only "dead, matured, or large growth of trees" could be harvested, and that these had to be individually "marked and designated." Since clearcuts take all the trees, regardless of size and without being specifically designated, environmentalists were able to persuade the federal courts to prohibit clearcutting in West Virginia's Monongahela National Forest. To prevent this ruling from being applied to the valuable Douglas-fir stands of the Pacific Northwest, the timber industry immediately pushed for new legislation which would supersede the Organic Act. The resulting NFMA did in fact permit clearcutting, but with certain restrictions. Where practiced, clearcutting had to be the "optimum" method of harvesting; it had to be shaped to match the terrain; and it had to be consistent with "the protection of soil, watershed, fish, wildlife, recreation, and esthetic resources." See C. F. Wilkinson, *Crossing the Next Meridian,* and Charles F. Wilkinson and H. Michael Anderson, *Land and Resource Planning in the National Forests* (Washington, D.C.: Island Press, 1987).

6. Ibid.

7. See above, Chapter 7.

8. Gordon Robinson, 1979: personal communication.

9. C. F. Wilkinson and H. M. Anderson, *Land and Resource Planning,* p. 183.

10. Ibid., pp. 124–125.

11. Tom Ribe, "Rangering from Capitol Hill," *Inner Voice* 3(2): Spring 1991.

12. C. F. Wilkinson and H. M. Anderson, *Land and Resource Planning,* p. 127.

13. Letter reprinted in *Inner Voice* 2(4): Summer/Fall 1990, p. 14.

14. Cited in C. F. Wilkinson, *Crossing the Next Meridian,* p. 152.

15. Letter reprinted in *Inner Voice* 3(5): Fall 1991, p. 1.

16. Letter reprinted in *Inner Voice* 2(4): Summer/Fall 1990, p. 15. The phrase "committment to meeting the ASQ" appears two more times in this letter.

17. Cited in Tom Ribe, "Pork Barreling Our National Forests," *Inner Voice* 3(2): Spring 1991, p. 5.

18. Tom Kovalicky, "Fighting Political Manipulation," *Protecting Integrity and Ethics: A Conference for Government Employees of Environmental, Wildlife and Natural Resource Agencies* (Eugene and Washington, D.C.: Government Accountability Project and Association of Forest Service Employees for Environmental Ethics, 1992), p. 6.

19. Randal O'Toole, "The Bottom Line: Congressional Pork Barrel," *Inner Voice* 2(2): Spring 1990.

20. Randal O'Toole, *Reforming the Forest Service* (Washington, D.C. and Covelo, CA: Island Press, 1988), p. 132.

21. Ibid., p. 134.

22. Ibid., p. 133.

23. Jim Wiebush, "Bureaucratic Gridlock Demoralizing USFS," *Inner Voice* 2(4): Summer/Fall 1990, p. 7.

24. Cheri Brooks, "Cutting from Below: Downsizing or Upsizing the Agency?" *Inner Voice* 5(1): Jan/Feb 1993.

25. Glen O. Robinson, *The Forest Service* (Baltimore: Johns Hopkins, 1975), p. 39.

26. P. J. Ryan, "Jousting with the Agency," *Protecting Integrity and Ethics,* p. 16.

27. B. W. Twight and F. J. Lyden, "Measuring Forest Service Bias," *Journal of Forestry* 87 (1989): 35–41.

28. G. O. Robinson, *Forest Service,* p. 34.

29. Cited in David A. Clary, *Timber and the Forest Service* (Lawrence: University of Kansas, 1986), p. 172.

30. C. F. Wilkinson, *Crossing the Next Meridian,* p. 162–3. "The problem here," Dwyer wrote, "has not been any shortcoming in the laws, but simply a refusal of administrative agencies to comply with them."

31. Ibid., p. 150.

32. R. G. Bennett, "Below-cost Sales: A Concern that Just Doesn't Ring True," *Below Cost Timber Sales, Conference Proceedings,* D. C. Le Master, B. R. Flamm, J. C. Hendee, eds. (Washington, D. C.: The Wilderness Society, 1987), p. 10.

33. Jim Kennedy, "Looking at Agency Values," *Protecting Integrity and Ethics,* p. 20.

34. Personal communication, 1993.

35. Hal Salwasser, "New Realities in the Forest Service," *Inner Voice* 4(4): July/August, 1992.

36. Jim Britell, "The Failure of Shasta Costa: How the Agency's New Perspectives Flagship Ran Aground," *Inner Voice* 4(4): July/ August, 1992.

Chapter 9

The interview with Gorden Tosten was conducted in 1980; the interviews with Wayne Miller and Henry Vaux were conducted in 1993.

1. R. W. Haynes, *Timber Situation in the United States.*

2. Ibid., p. 50.

3. U.S.D.A. Forest Service, *The Outlook for Timber in the United States, Report of the Findings of the 1970 Timber Review.* Cited in Barney, *Last Stand,* p. 34.

4. M. Dodge, et al., *Erosion Rates,* pp. 83–88.

5. *California Forest Practice Rules* (Sacramento: California Department of Forestry, 1992), p. 42. Emphasis added.

6. D. P. Gasser, "Impacts of Forest Practice Regulations in California," *Forest Operations in Politically and Environmentally Sensitive Areas* (Tahoe City: Council on Forest Engineering, 1985), 11–15.

7. M. Gravelle, "Three Acre Loophole: Logging for Dollars in Your Own Backyard," *North Coast Journal* 4(7): July, 1993, pp. 18–20.

Chapter 10

The interviews with Randy Spanfellner and Walter Smith were conducted in 1993.

1. R. W. Haynes, *Timber Situation in the United States,* p. 64.

2. E. C. Weeks, "Mill Closures in the Pacific Northwest: The Consequences of Economic Decline in Rural Industrial Communities," *Community and Forestry: Continuities in the Sociology of Natural Resources.* R. G. Lee, D. R. Field, and W. R. Burch, Jr., eds. (Boulder, San Francisco, and London: Westview Press, 1990), pp. 125–140.

3. Ibid.

4. Ibid.

5. A. Brunclle, "The Changing Structure of the Forest Industry in the Pacific Northwest," *Community and Forestry.* R. G. Lee *et al.,* eds., pp. 107–124.

6. C. H. Schallau, "Community Stability: Issues, Institutions, and Instruments," *Community and Forestry.* R. G. Lee *et al.,* eds., pp. 69–82.

7. *Final Environmental Impact Statement on Management for the Northern Spotted Owl in the National Forests* (Washington, D. C.: USDA Forest Service, 1992), p. 127.

8. V. A. Sample and D. C. Le Master, "Economic Effects of Northern Spotted Owl Protection." *Journal of Forestry* 90 (1992): 31–35. This article reviews and compares the estimates of employment impacts made in four major reports. Although the range appears to vary from 12,000 to 147,000, Sample and Le Master find that much of this apparant variation is caused by the use of differing variables.

9. Ibid.

10. E. C. Weeks, "Mill Closures," *Community and Forestry.* R. G. Lee *et al.,* eds.

11. V. A. Sample and D. C. Le Master, "Economic Effects of Northern Spotted Owl Protection."

12. *Report of the Society of American Foresters National Task Force on Community Stability* (Bethesda, MD: Society of American Foresters, 1989).

13. William Stewart, "Trends in Labor Productivity Gains in the California Timber Industry," (Unpublished paper, Department of Forestry and Resource Management, University of California at Berkeley). Figures are derived from the California Economic Development Department.

14. H. M. Anderson and J. T. Olsen, *Federal Forests and the Economic Base of the Pacific Northwest* (Washington, D.C.: The Wilderness Society, 1991); J. H. Beuter, *Social and Economic Impacts of the Spotted Owl Conservation Strategy* (Washington, D.C.: American Forest Resources Alliance, 1990).

15. C. H. Schallau, "Community Stability," *Community and Forestry.* R. G. Lee *et al.,* eds.

16. H. M. Anderson and J. T. Olsen, *Federal Forests and the Economic Base.*

17. Wecks, "Mill Closures," *Community and Forestry.* R.G. Lee *et al.,* eds.

18. Jonathan Kusel and Louise Fortmann, *Well-Being in Forest-Dependent Communities* (Berkeley: Department of Forestry and Resource Management, University of California, 1991), p. xii.

19. V. A. Sample and D. C. Le Master, "Economic Effects of Northern Spotted Owl Protection."

20. W. Stewart, "Labor Productivity Gains in California."

21. P. Bettinger, J. Sessions, and L. Kellogg, "Potential Timber Availability for Mechanized Harvesting in Oregon," *Western Journal of Applied Forestry* 8 (1993): 11–15.

22. V. A. Sample and D. C. Le Master, "Economic Effects of Northern Spotted Owl Protection."

23. R. W. Haynes, *Timber Situation in the United States,* p. 55.

24. H. M. Anderson and J. T. Olsen, *Federal Forests and the Economic Base.*

25. D. F. Flora and W. J. McGinnis, "An Analysis of the Effects of Northern Spotted Owl Conservation, Harvest Replanning, a Log Embargo, and Recession of the Northwest Log and Lumber Trade," *Western Journal of Applied Forestry* 6(1991): 87–89.

26. V. A. Sample and D. C. Le Master, "Economic Effects of Northern Spotted Owl Protection," p. 35.

Chapter 11

The interview with Dr. Rudolph Becking was conducted in 1980; the interviews with Nadine Bailey, Joseph Bower, Bob Ford, Leah Wills, and Mark Harmon were conducted in 1993.

1. R. A. Sedjo, "International Competitiveness, Community Stability and Adapting to a Rapidly Changing Global Economy," *Community Stability in Forest-Based Economies.* D. C. Le Master and J. H. Beuter, eds. (Portland: Timber Press, 1989), p. 180.

2. Chris Maser, "The Future is Today: For Ecologically Sustainable Forestry," *Inner Voice* 3(5): Fall 1991, p. 12.

3. J. F. Franklin and C. T. Dyrness, *Natural Vegetation of Oregon and Washington,* General Technical Report PNW-8 (Portland: Pacific Northwest Forest and Range Experiment Station, USDA Forest Service, 1973).

Glossary

A-Horizon Scientific term for the upper layer of soil.

AFSEEE Association of Forest Service Employees for Environmental Ethics.

All-Age Management (also, *All-Aged Management*) The management of a forest to include trees of diverse ages and sizes.

Allowable Cut Effect (ACE) See *Earned Harvest Effect.*

Allowable Sale Quantity (ASQ) Technically, the maximum harvesting level for the National Forests; in practice, a target level for harvesting.

Anadromous Fish Ocean fish that migrate up a river to spawn.

Angle of Repose The maximum slope at which a hillside will remain stable; a hill that is made steeper than its angle of repose is likely to crumble.

Back Hoe A piece of earth-moving equipment with a trench-digging shovel on one end and a loading scoop on the other.

Bare-Root Seedlings Seedlings germinated together in a single mass of dirt instead of in their own separate containers.

Below Cost Sales (BCS) Government timber sales in which the receipts are less than the costs of administering the sales.

Biodiversity Biological diversity; a wide variety of living organisms.

Biomass The sum total of organic material in a given area.

BLM The Bureau of Land Management, an agency of the United States Department of the Interior.

Board Foot A unit of measurement nominally equal to a board measuring 1 in. × 12 in. × 12 in.

Boomer A migrant logger.

Broadcast Burn Controlled fire over the entire surface of a designated area.

Broadleaf Tree A tree with conventional leaves, not needles or scales.

Buck To cut a downed tree into logs of specified lengths.

Bucker A person who cuts downed trees into logs.

Bull Donkey A large steam donkey operating along skid roads.

Bull-Puncher (also, *Bull-Whacker*) The driver of a logging ox-team.

Butt Log The log taken from the base of a tree, often slightly irregular.

Calk Boots Logging boots with short spikes set in the soles.

Canopy The upper level of vegetation in a forest.

Cat Abbreviation for caterpillar.

Caterpillar In common usage, a logging tractor. (In this text, the word *caterpillar* denotes any logging tractor; the word *Caterpillar* denotes a logging tractor made by the Caterpillar Company.)

Catskinner The operator of a logging tractor.

CDF California Department of Forestry.

Chaser A person who works on the landing in a cable operation, unhooking logs from the chokers.

311

Check Dams A series of small dams in a stream or drainage ditch, used to minimize erosion.

Choker A cable loop that is attached to a log during the yarding of timber.

Choker Setter A person who puts chokers around logs.

Clearcutting The removal of the entire stand of trees within a designated area.

Climax The culminating stage of plant succession for a given environment, where the vegetation has reached a stable condition.

Codominant Trees Trees that reach into the upper forest canopy but do not extend above their neighbors.

Conifer A cone-bearing tree that generally has needles or scales. Most commercial tree species in the Northwest are conifers.

Conk Fruit-mass of a wood-destroying fungus, generally projecting from the trunk of a tree.

Contour Wattles A series of hillside terraces supported by a network of live shoots and intertwined dead branches.

Cord A unit of measurement for firewood or pulpwood, equal to a pile measuring 4 ft × 4 ft × 8 ft, or 128 cubic feet.

Corks See *Calk Boots*.

Crawler Tractor A logging tractor with ribbed metal tracks instead of wheels.

Cruise A survey to measure the timber in a designated area.

Crummy A truck-like van used to transport workers.

Cull A tree or log that is rejected because of defects.

Culmination of Mean Annual Increment (CMAI) The time at which the average growth of a tree is maximized; a measure of productive maturity.

d.b.h. Diameter at breast height (4.5 feet), a standard method for determining the size of a tree.

Deciduous Forest A forest primarily comprising trees that shed their leaves in the winter.

Deck A pile of logs; logs that are about to be transported or milled form a hot deck, while logs that are being stored form a cold deck.

Dibble A pointed digging tool used for planting containerized seedlings.

Discount Rate See *Guiding Rate of Interest*.

Dog A hook.

Dog-Hole Port A small, often tenuous, anchoring site along a rugged coastline.

Dominant Trees Trees that are taller than their immediate neighbors.

Donkey Puncher The operator of a steam donkey.

Duff The uppermost layer of the forest floor, composed of decomposing organic material.

Earned Harvest Effect (EHE) The increase in allowable harvests because of projected gains in future growth due to intensive management.

Economic Maturity The point at which more money can be made by harvesting timber than by letting it grow; at economic maturity, a new investment will be more financially lucrative than a continuation of the original investment.

Ecosystem The complex of interactions between a community of living organisms and their environment.

Ecosystem Management Forest management based on the goal of maintaining a healthy ecosystem.

EIR, EIS Environmental Impact Report, Environmental Impact Statement; studies that evaluate the effects of a proposed action on the environment.

EPA Environmental Protection Agency, an arm of the federal government charged with evaluating the safety of commercially used chemicals.

Erosion Hazard Rating A number, based on the soil type and the steepness of the slope, which is supposed to determine a site's susceptibility to erosion.

Even-Age Management (also, *Even-Aged Management*) Forest management in which the crop trees are all of a similar age.

Fall (also, *Fell*) To cut down a tree.

Faller A person who cuts down trees.

Feller-Buncher A machine that cuts trees with giant shears and then stacks the trees in a pile.

FIP Forestry Incentives Program, a federal cost-sharing program that helps small timber owners pay for reforestation or timber stand improvement.

FORPLAN A computer program used by the Forest Service to calculate economic value.

Fry Young fish, just after they are hatched.

Garlon Brand name for triclopyr, an herbicide.

Girdling Killing a tree by removing a strip of bark from around its trunk.

Grapples Giant arms on a logging machine which can grab logs and move them from place to place.

Green Chain An assembly-line step in mill work; the sorting of lumber after it has run through the saw.

Green Logging The logging of timber that is still alive; the opposite of salvage logging.

Green-Manure Trees Trees that fertilize the ground as they grow, generally by the presence of nitrogen-fixing bacteria living on their roots.

Greenwashing The practice of using environmental slogans in the advertising of nonenvironmental products.

Guiding Rate of Interest The rate of interest that is expected to prevail throughout the economy during a specified period.

Guy Line A support cable for a spar tree or yarding tower.

Gyppo Logger A small, independent contractor.

Hand Briar See *Misery Whip*.

Hardwood Popular term for a broadleaf tree.

Haulback Line The cable that returns the chokers to the woods after the yarding of a turn of logs.

Haul Road A major road engineered for logging trucks as well as tractors.

Heartwood The inner layers of a tree that have ceased to contain living cells and which serve only for structural support; lumber coming from heartwood is generally stronger and rots less quickly than that coming from the outer layers of the tree (sapwood).

Herbicide A chemical used to control unwanted vegetation.

High-Grading Logging in which the high-quality timber is removed while the low-quality timber is left standing.

High-Lead Cable logging in which the main block is suspended from a spar tree or yarding tower so that the front end of the log can be lifted off the ground.

Hoedad A digging tool used for planting seedlings.

Holistic Forestry An approach to forestry which treats the ecosystem as a self-equilibrating mechanism and which proposes only minimal, site-specific interference with natural processes.

Hooker   In modern usage, a person who attaches the tag lines dangling from a helicopter to the chokers attached to the logs; in the old days, another word for hooktender.

Hooktender   Woods boss, or foreman of a yarding crew.

Ingrowth   An increase in the value of timber caused by the higher price which can be obtained from products of older trees.

Intensive Management   The entire complex of industrial tree-farming techniques: genetic selection, site preparation, planting, weed control, pest control, fertilization, thinning, harvesting.

Internal Rate of Return (IRR)   The compounded annual interest rate earned on the initial investment.

Intolerant   Most often, short for shade intolerant, referring to the inability of certain trees to thrive under a forest canopy.

Kerf (also, *Curf*)   The width of a saw cut, determining the amount of wood which is turned into sawdust instead of lumber.

Knutsen–Vandenberg (K–V) Funds   A share of the timber sale receipts which can be retained by National Forest districts for reforestation and related activities.

Landing   A place where logs are assembled for loading.

Leader   The top of a tree, representing the most recent growth.

Mainline   The cable that moves the logs in a yarding operation.

Manual Release   Hand-clearing of brush (as contrasted with the use of herbicides) to lessen the competition around desired trees.

MBF   One thousand board feet.

Mean Annual Increment (MAI)   The average annual growth rate for a tree, computed over its entire life cycle.

Mensuration   The branch of forestry concerned with the measurement of timber.

Millrace   The controlled channel of water leading up to a mill.

Misery Whip   A large, hand-operated crosscut saw used for falling and bucking timber.

MMBF   One million board feet.

Monoculture   The raising of a tree crop that consists of a single species.

Multiple Use   Management that encourages several distinct uses of the forest.

MUSY   Multiple Use–Sustained Yield Act of 1960.

Mycorrhizae   Fungi that grow among the outer cells of plant rootlets and form a symbiotic relationship with their hosts. The fungi receive photosynthetic products such as carbohydrates and vitamins from the host plants; in return, they help their hosts absorb nutrients and water from the soil.

NEPA   National Environmental Policy Act of 1970.

New Forestry   See *Holistic Forestry*.

NFMA   National Forest Management Act of 1976.

Nitrogen Fixation   The transformation of atmospheric nitrogen into nitrogen compounds that can be used by growing plants.

Nondeclining Even Flow (NDEF)   A Forest Service harvesting standard requiring that current yields have to be sustainable even after there is no more old-growth timber to cut.

Nonindustrial Timber Management Plan (NTMP)   In California, a long-term plan for small landowners which can substitute for separate timber harvest plans.

Old Growth   According to the timber industry, old and large trees; according to biologists, an ecosystem which includes mature trees, a multilayered canopy, snags, large woody debris, and distinctive wildlife habitat; according to preservationists, a forest which has never been logged.

Overstory See *Canopy.*

Overstory Removal The harvesting of the tallest trees without purposely removing the smaller vegetation.

Overtop To grow above the neighboring vegetation.

Peavy A wooden pole, with a spike at the end and a hinged hook, used for moving and turning logs by hand.

Peeler A log of sufficient size and quality for making rotary-cut veneer.

Pesticide A chemical used to kill pests which prey on crop trees.

Phenoxy Herbicides A group of closely related chemical compounds (most commonly 2,4-D; 2,4,5-T; and 2,4,5-TP) which can kill, or severely damage, many species of broadleaf plants by promoting the uncontrolled expansion and division of cells.

Photosynthesis A green plant's transformation of carbon dioxide and water into carbohydrates, using sunlight as the source of energy.

Pioneer A plant belonging to the earliest stage of vegetational succession; pioneers establish themselves quickly in the wake of cataclysmic changes and then are gradually replaced by successor species.

Plus Tree A tree thought to be of superior genetic quality.

Present Net Worth (PNW) The economic value of a given timber site (including the current stand), taking into account all projected revenues and costs.

Productivity Class See *Site Class.*

Productive Maturity The point at which a new crop will produce more timber than the original crop.

Release Freeing the crop tree from immediate competition by eliminating, or retarding the growth of, the neighboring vegetation.

Riffle A shallow rapid in a stream.

Rigging The cable network, including blocks and related hardware, used in yarding timber or in directional falling.

Rigging Slinger A person who attaches chokers to the main yarding cable.

Riparian Having to do with a river or stream environment.

River Rat (also, *River Hog*) A person who follows the logs on a river run and pries them loose when they cease to move.

Rotation The number of years allotted to a single tree crop in a given location.

RPA Resources Planning Act of 1974.

Salvage Logging Logging in which only dead timber (either standing or on the ground) is removed; the opposite of green logging.

Sapwood The outer layers of a tree that contain living cells; lumber coming from sapwood is generally weaker and rots more quickly than that coming from the inner layers of the tree (heartwood).

Scale To measure harvested logs.

Second Growth Young trees that grow in the wake of a prior harvest.

Seed Tree A tree that is purposely left standing during a logging operation in order to produce natural regeneration in the surrounding area.

Selective Cutting The periodic harvesting of selected trees scattered throughout a forest, either individually or in small groups.

Shelterwood Cutting The harvesting of timber in two or more successive stages, where the trees left standing after the first harvest provide a natural seed source and partial protection for regeneration.

Show A logging operation.

Side A complete yarding and loading crew.

Side-Spooler A single-engine, single-spool steam donkey.

Silviculture The cultivation of forest trees.

Site Class A grouping of similar site indices.

Site Index A measure of the productive capacity of a given area, based on the height of the dominant trees at a specified age; a site index of 170 for Douglas-fir means that a 100-year-old dominant tree in that stand will be about 170 feet tall.

Site Preparation Preparing an area for planting, often with the aid of controlled fires, herbicides, or mechanized devices.

Skid To drag logs along the ground from the forest to a landing.

Skid Road A road used for skidding timber. In the old days, skid roads were made of wood; today, a logging tractor carves its own skid roads in the dirt before removing the timber.

Skid Trail A path through the woods created by the skidding of logs.

Skyline A cable logging system in which the logs are suspended in air rather than dragged along the ground.

Slacker An early term for a skyline cable.

Snag A standing dead tree.

Softwood Popular term for conifer.

Soil Expectation Value (SEV) An economic measure of the capacity of unstocked land to produce timber.

Spar Tree A tree that has been limbed, topped, and rigged with cables to provide elevation in high-lead yarding.

Species A category of organisms which can reproduce with each other but not with other types of organisms.

Splash Dam A temporary dam on a shallow stream. When the water behind the dam is released, logs can be propelled downstream.

Springboard A board inserted into a notch in the trunk of a large tree, forming a one-piece scaffolding upon which a faller can stand while he cuts down the tree.

Stand A community of trees managed as a collective unit.

Steam Donkey A steam-driven yarding machine used in the late nineteenth and early twentieth centuries.

Stocking The quantity of crop trees in a designated area, often expressed as a percent of maximum capacity.

Strip-Cutting The harvesting of parallel strips of timber, in which the areas between those strips are left untouched.

Stumpage The quantity or value of standing timber.

Succession The replacement of one plant community by another as local conditions change over time.

Successor Species Species of plants that thrive in the intermediate stages of forest succession, after the pioneers but before the climax stage.

Sunscald Damage done to a tree because of sudden and excessive exposure to the sun.

Super Trees The timber industry's term for fast-growing, high-yield trees created by genetic engineering and aided by intensive forest management.

Suppressed Tree A tree that remains in the forest understory because its growth is inhibited by neighboring vegetation.

Sustained Yield The amount of timber that a forest can produce on a continuing basis in perpetuity.

Swing Donkey A steam donkey used to supplement another steam donkey over a long haul.

Tag Line A short stretch of cable that connects the chokers and logs with the main part of the yarding system.

Tailhold The anchor on the far end of a cable yarding system.

THP Timber Harvest Plan, a document that states how a timber harvest will be executed; by law, a THP must be filed with the State Department of Forestry prior to any logging operation in California.

Timber Stand Improvement The pruning, weeding, and/or thinning of a stand of timber.

Tolerance Most often, the ability of a tree to grow in the presence of shade; more generally, the ability of a tree to grow amid competition or under other adverse conditions.

Transpiration The process by which plants release water into the atmosphere.

Triclopyr An herbicide which is now more commonly used than phenoxy herbicides because it is believed to be safer.

Turn of Logs A group of logs yarded at the same time by the same machine or animal.

Understory Vegetation which is shaded by the forest canopy.

Uneven-Age Management (also, *Uneven-Aged Management*) See *All-Age Management.*

Water Bar A mound of dirt placed in a logging road to direct water away from the roadbed, thereby lessening the damage from erosion.

Watershed An area defined by a particular drainage system.

Waterslinger A person who attaches logs to a donkey-powered cable along a wooden skid road and then lubricates the skid road as the logs are dragged to the landing.

Whistle Punk A person who communicates signals to the various workers on a steam-donkey yarding crew.

Widow Maker A high, loose branch that presents a danger to loggers working beneath it.

Windthrow (also, *Windfall* or *Blowdown*) The destruction of standing timber by the wind.

Wobblies Members of the Industrial Workers of the World, a radical labor organization during the first quarter of the twentieth century.

Yarder A machine used in yarding timber.

Yarding The movement of logs from the woods to a central loading area, or landing.

Yield Table A table that attempts to predict how much timber of a given species could be grown in a perfectly stocked stand with a specified site index.

Index

"Mary Shomon is more knowledgeable and has provided patients with more timely and comprehensive information about these conditions than most physicians. As a result of Mary's tireless work, many patients previously undiagnosed or inadequately treated are now getting the necessary tests and treatment they need."

—Marie Savard, MD, author of *How to Save Your Own Life*

"Patients who are struggling to get diagnosed with or understand their autoimmune conditions will find Mary Shomon's book an extremely useful, essential resource to help them find the practitioners and treatment—both conventional and alternative—that will help them truly live well."—Stephen E. Langer, MD, author of *Solved: The Riddle of Illness*

"Mary Shomon is a leader in understanding the information patients need when it comes to autoimmune and thyroid disorders. Not only do patients need this information, but her books should be required reading by doctors so they understand the concerns and fears of their patients." —Larian Gillespie, MD, author of *You Don't Have to Live with Cystitis*

"The truth, the whole truth, and nothing but the truth about autoimmune diseases from Mary Shomon, the woman who taught America about thyroid disease! Her book will enlighten you about how to diagnose, treat, and possibly even prevent, autoimmune disease in yourself or someone you love. Mary has my highest respect for her careful research and the way she presents both the conventional and the alternative aspects in a way that anyone can understand." —Carol Roberts, MD, Director, Wellness Works Holistic Health Center

"*Living Well with Autoimmune Disease* is a much needed book. It gives hope to those who suffer from chronic illnesses. This is a wonderful book that provides information and solutions, and I would highly recommend this book to my patients."

—David Brownstein, MD, author of *The Miracle of Natural Hormones*